GROWING A SUSTAINABLE CITY?

The Question of Urban Agriculture

Urban agriculture offers promising solutions to many different urban problems, such as blighted vacant lots, food insecurity, storm water runoff, and unemployment. The objectives of urban agriculture connect to many cities' broader goal of "sustainability," but tensions among stakeholders have started to emerge as urban agriculture is incorporated into municipal policy-making frameworks.

Growing a Sustainable City? offers a critical analysis of the development of urban agriculture policies and their role in shaping post-industrial cities. Christina Rosan and Hamil Pearsall's case study of Philadelphia demonstrates how growing food in the city has become a symbol of urban economic revitalization, sustainability, and – increasingly – gentrification. Their research includes interviews with urban farmers, gardeners, and city officials, and reveals that the transition to "sustainability" is marked by a series of tensions along race, class, and generational lines. The book evaluates the role of urban agriculture in sustainability planning and policy by placing it within the context of a large North American city struggling to manage competing environmental and economic objectives. While highlighting the challenges and opportunities of institutionalizing urban agriculture into formal city policy, *Growing a Sustainable City?* tells the story of a city's growing pains as it attempts to reinvent itself as a green, attractive, and prosperous place to live.

(UTP Insights)

CHRISTINA D. ROSAN is an associate professor in the Department of Geography and Urban Studies at Temple University.

HAMIL PEARSALL is an assistant professor in the Department of Geography and Urban Studies at Temple University.

UTP insights

UTP Insights is an innovative collection of brief books offering accessible introductions to the ideas that shape our world. Each volume in the series focuses on a contemporary issue, offering a fresh perspective anchored in scholarship. Spanning a broad range of disciplines in the social sciences and humanities, the books in the UTP Insights series contribute to public discourse and debate and provide a valuable resource for instructors and students.

BOOKS IN THE SERIES

GROWING A SUSTAINABLE CITY?

The Question of Urban Agriculture

Christina D. Rosan and Hamil Pearsall

UNIVERSITY OF TORONTO PRESS
Toronto Buffalo London

© University of Toronto Press 2017
Toronto Buffalo London
www.utorontopress.com
Printed and bound by CPI Group (UK) Ltd, Croydon, CR0 4YY

ISBN 978-1-4426-5061-9 (cloth)   ISBN 978-1-4426-2855-7 (paper)

Printed on acid-free, 100% post-consumer recycled paper with vegetable-based inks.

Library and Archives Canada Cataloguing in Publication

Rosan, Christina, author
Growing a sustainable city? : the question of urban agriculture / Christina D. Rosan
and Hamil Pearsall.

(UTP insights)
Includes bibliographical references and index.
ISBN 978-1-4426-5061-9 (hardcover). – ISBN 978-1-4426-2855-7 (softcover)

1. Urban agriculture.   2. Urban agriculture – Pennsylvania – Philadelphia – Case
studies.   3. Sustainable agriculture.   4. Sustainable agriculture – Pennsylvania –
Philadelphia – Case studies.   5. Sustainable urban development.   6. Sustainable
urban development – Pennsylvania – Philadelphia – Case studies.   7. City planning.
8. City planning – Pennsylvania – Philadelphia – Case studies.   I. Pearsall, Hamil,
1981–, author   II. Title.   III. Series: UTP insights

S494.5.U72R67 2017   630.9173'2   C2017-904603-9

University of Toronto Press acknowledges the financial assistance to its publishing
program of the Canada Council for the Arts and the Ontario Arts Council, an agency
of the Government of Ontario.

 Canada Council   Conseil des Arts
for the Arts   du Canada

ONTARIO ARTS COUNCIL
CONSEIL DES ARTS DE L'ONTARIO
an Ontario government agency
un organisme du gouvernement de l'Ontario

Funded by the   Financé par le
Government gouvernement
of Canada   du Canada   Canadä

For Robert Mason

Contents

Contents

Figures

GROWING A SUSTAINABLE CITY?

The Question of Urban Agriculture

chapter one

Introduction:
The Dilemma of Urban Agriculture

In Philadelphia, a city that two-term mayor Michael Nutter promised to make "the Greenest City in America" when he was inaugurated in 2008, the excitement over urban agriculture is ubiquitous.[1] A 2011 *Philadelphia Inquirer* article titled "City's New Wave of Farmers" features interviews with young urban farmers, many of whom moved to Philadelphia because they "heard about the scene" (Marder 2011). *Grid Magazine*, Philadelphia's local sustainability magazine, routinely features articles on local and sustainable foods and urban agriculture. Philadelphia is an urban agriculture hotspot: people are raising bees on their rooftops, advocating for backyard chickens at city meetings, collecting rainwater in rain barrels, guerrilla gardening on vacant lots, eating hyper-local and organic food, teaching students about growing, selling, and cooking healthy food, shopping at farmers' markets, composting, and mobilizing political support to garden and farm in the city. Community gardens are located all across the city. Entrepreneurs are starting farms on vacant land and experimenting with commercial vertical farms. Undergraduate students are learning about urban agriculture, hoop houses, aquaponics, and how to use urban agriculture as a means of transforming an unsustainable food system. Some of them want to be professional urban farmers when they graduate. There is a sense among these enthusiasts that urban agriculture is one of the keys to making Philadelphia the "Greenest City in America" (which is Target 15 in the city's sustainability plan).[2] They are excited about the possibilities.

The urban agriculture bandwagon in cities is full steam ahead, and urban agriculture is increasingly viewed as an integral component of a sustainable city. This is true for global cities like San Francisco, New York, Vancouver, and London as well as post-industrial cities like Detroit, Cleveland, and Philadelphia that are still working to redefine themselves. Activists who believe in the transformative power of urban agriculture are making their voices heard to ensure that it is connected to the urban sustainability agenda. And the promises are big. It is difficult to attend an urban sustainability event in Philadelphia without hearing about the potential of urban agriculture for addressing blight, revitalizing communities, improving food security in underserved neighbourhoods, mitigating stormwater run-off, and providing valuable youth engagement opportunities. Much of the rhetoric lauds urban agriculture and the benefits it brings. A growing body of academic literature on urban agriculture has also supported this positive view of it (Draper and Freedman 2010). These studies tout the mental, nutritional, and physical health outcomes of urban agriculture, and the opportunities it provides for youth engagement, besides pointing to the ways in which it promotes social capital (Alaimo et al. 2008; D'Abundo and Carden 2008; Shinew, Glover, and Parry 2004; Blair 2009; Glover, Shinew, and Parry 2005).

Scholars, planners, and policy-makers have embraced the potential of urban agriculture in its many forms (e.g., community gardens, urban farms), viewing it as a mechanism for promoting long-term economic development, community revitalization, and social and environmental health. For instance, Wachter's (2004) study found that urban greening and vacant land stabilization can increase surrounding property values by up to 30 per cent. In some places, community gardens increased property values within 1,000 feet, particularly over time and in disadvantaged neighbourhoods (Voicu and Been 2008). With regard to food security, urban agriculture can provide nutritious food to low-income communities, thus addressing a challenge that many cities face (e.g., Armstrong 2000). In terms of social capital, urban agriculture can engage community residents and promote community organizing (e.g., Firth, Maye, and Pearson 2011), provide educational and employment opportunities for low-income youth (e.g., Ferris, Norman, and Sempik 2001), increase green space in the city, play a role in retaining stormwater and mitigating water pollution (e.g., Lovell and Johnston 2009), and "restore the natural environment" (e.g., Irvine, Johnson, and Peters 1999, 34).

While much of the research on urban agriculture focuses on its benefits, some scholars have brought a critical perspective to urban gardening as well as to alternative food movements that seek to address inequities in food systems and food accessibility through farmers' markets, community gardens, and other ways to deliver food to underserved communities (Alkon and Agyeman 2011). This line of critical research does the important work of illuminating the politics of race and class – a politics that is often ignored and that may make urban agriculture enthusiasts uncomfortable. Urban gardens and farms, which are often celebrated as efforts to bring fresh produce to underserved, largely minority communities, are often led by middle-class whites and thus may reflect white values and fail to speak to the communities these gardening projects purport to serve (Guthman 2008; Slocum 2007). In short, while the stated intent of many urban agriculture initiatives may be to support low-income and minority communities, in reality, the complicated politics of race and class may well limit the potential of alternative food movements to meaningfully redress urban food inequities (Bedore 2010).

These critical food scholars have focused on the politics of urban food production; less attention has been devoted to the land use politics surrounding the formalization of alternative food provision – specifically urban agriculture – within planning and policy-making. We argue that this is an important area of inquiry that needs to be more closely examined since cities around the world are grappling with how to integrate urban agriculture into their policy and planning frameworks. As of the early 2000s, only a few cities in North America had institutionalized urban agriculture as a long-term policy solution (Lawson 2005), but this is changing, and urban agriculture is becoming part of the larger conversation about the sustainable redevelopment of cities. Municipal leaders are including urban agriculture in official planning documents and policies as a way to stabilize communities and promote community, economic, and environmental benefits. Planners no longer relegate urban agriculture to the category of a hobby or eccentric use of urban space; it has become increasingly accepted as a low-cost approach to addressing a host of urban challenges and promoting sustainability (Thibert 2012). It is also seen as a desirable amenity that may help attract what Richard Florida calls "the creative class," a demographic of skilled professionals who are seen as critical to urban vibrancy and economic development (Florida 2014). In fact, the American Planning Association (APA) has issued numerous reports on urban agriculture and food systems planning, having recognized that both fall squarely under the purview of planning and economic development (Mukherji and Morales 2010; Hodgson, Campbell, and Bailkey 2011). But despite the growing acknowledgment of urban agriculture as an accepted land use, figuring out how to formally integrate urban agriculture into planning and urban development policies, and how to do so while addressing issues of equity and social justice, is a relatively new task.

This book critically examines the evolution of urban agriculture in Philadelphia's city land policies and the motivations of those who advocate for formalizing urban agriculture into planning. Our study builds on previous scholarship (e.g., Bassett 1979; Lawson 2005) that traces the ebbs and flows of urban agriculture activities through history. We contribute to this history by linking the current interest and policy developments in urban agriculture and sustainability to Philadelphia's history of urban gardening and growing. We seek to explain how urban agriculture has become a component of the city's sustainability policy as it attempts to reinvent itself as a green, attractive, and growing city recovering from decades of poverty, large-scale vacancy and abandonment, racial tensions, unemployment, a failing school system, and high rates of crime.

We explore the politics surrounding the institutionalization of urban agriculture and how this process has included or excluded various perspectives and stakeholders, who include long-time gardeners, new gardeners, and community leaders. By focusing on policy-making, we are able to describe the ongoing tensions surrounding the role played by urban agriculture in a city that after decades of economic depression and abandonment has begun to attract new development attention and gentrification. We seek to uncover the process of policy creation around urban agriculture and what it means for Philadelphia's diverse constituents.

The Dilemma of Urban Agriculture in Philadelphia

Philadelphia is a city in transition, one where optimism and youthful enthusiasm confront daunting urban challenges. Philadelphia lost one-quarter of its population between 1950 and 2000 and currently has an estimated 40,000 vacant lots (Figure 1).[3] City officials are faced with the daily challenges of managing a city that Pew Charitable Trusts described in 2014 as "the poorest big city in America," one that "also has the highest rate of deep poverty – people with incomes below half of the poverty line – of any of the nation's 10 most populous cities. Philadelphia's deep-poverty rate is 12.2 percent, or nearly 185,000 people, including about 60,000 children. That's almost twice the U.S. deep-poverty rate of 6.3 percent."[4]

Even while continuing to suffer a 26 per cent poverty rate, Philadelphia is attracting new residents.[5] And building permits are up: the city has grown by 78,732 residents since 2006, and many of these newcomers are millennials.[6] The *Pew State of the City* report for 2016 sums up the challenge Philadelphia is facing as it wrestles with recent demands for development: "This is a lot of change in a relatively short time. But other aspects of life in Philadelphia have not changed – at least not enough to make much of a difference."[7] The "other aspects" that the Pew report refers to are challenges that continue to plague Philadelphia such as high crime and poverty rates, failing schools, and vacant land, particularly in low-income, primarily minority communities (Figure 1).[8] According to the city's *Land Bank: 2017 Strategic Plan and Performance Report 2017*, while parts of the city are rapidly redeveloping, "forty percent of Philadelphia's population, or 594,040 people, reside in a stressed or distressed market where the median sales price for homes is less than $50,000 and vacancy levels are at the highest rates."[9]

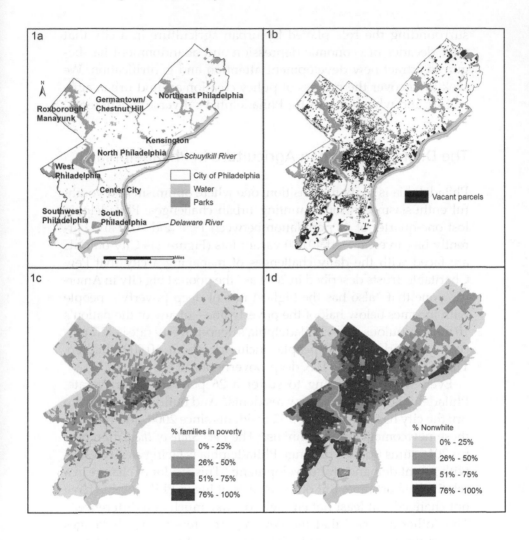

Figure 1a. General geography of the City of Philadelphia. 1b. Vacant lots, based on 2016 Land Use data from the City of Philadelphia. 1c. Percentage of families in poverty, based on 2009–13 American Community Survey estimates. 1d. Percentage of non-white families, based on 2009–13 American Community Survey estimates. Sources: City of Philadelphia; 2009–13, 5-year American Community Survey (US Census).

Today, many Philadelphia city officials anticipate *growth* – in population and real estate development – and urban greening, through green infrastructure and urban agriculture, is an important part of their vision. When Mayor Nutter (2008–16) took office, he saw an opportunity in the growing sustainability movement. He declared that Philadelphia would become the "Greenest City in America," and he committed to this vision by creating the Mayor's Office of Sustainability.[10] In 2009 Philadelphia released its first municipal sustainability plan, *Greenworks*, to serve as a central visioning document for the sustainable development of the city. *Greenworks* is a sustainability plan that anticipates the growth and revitalization of Philadelphia. The plan recognizes Philadelphia's history of economic decline and abandonment; however, it describes the city's 40,000 vacant lots as "assets" rather than liabilities, and its stated goal is to turn Philadelphia into a sustainable and prosperous city.[11] The hope of redevelopment that underscores *Greenworks* is echoed in the city's subsequent policy documents, specifically, the Land Bank report: "parcels that appear to be blighted sites today can become locations for infill housing and commercial development to meet the growing demand from existing and new households, particularly as we try to seek ways to support economic opportunity for Philadelphia's low- and moderate-income households."[12] This report, like *Greenworks*, frames vacancy as opportunity.

Where does urban agriculture fit into the *Greenworks* plan and the city's goal of being sustainable? Philadelphia has a long history of community gardening and urban agriculture that is often ignored or underplayed. For decades, city officials regarded urban agriculture as a logical policy approach: as a low-cost, high-rewards way to anchor communities (by demonstrating the investment of social and economic capital), one that put "eyes on the street," that cleaned and greened, that promoted economic development, and that ultimately grew the city's tax base as redevelopment happened around these stabilized lots.[13] Arguably, though, a shift occurred when the city released *Greenworks*, because with that document, the city was committing to a more formal effort to include urban agriculture in city policy-making. Urban agriculture has moved from an activity supported by non-profits and tacitly acknowledged (and sometimes purposely ignored) by the government to an activity the city recognizes and supports.

One challenge facing urban agriculture activists and scholars in Philadelphia is quantifying the actual number of urban gardens and farms in the city. This process is contested and political. Domenic Vitiello and Michael Nairn in 2008 estimated that the number of community gardens in Philadelphia had actually declined, from 501 to 226 between 1996 and 2008 (Vitiello and Nairn 2009, 3). According to the city's *Greenworks 2015 Progress Report*, there were 343 "gardens, markets, and farms" in the city, up from 230 in 2008.[14] The Garden Justice Legal Initiative (GJLI), a legal advocacy organization working to protect urban gardens and farms, put the number of current gardens and farms in Philadelphia much higher, at between 400 and 500.[15] While we do not know the exact number of gardens and farms today, we do know that they are an important part of the urban fabric of Philadelphia and that they have been for generations. They are spaces where people come together to grow food, socialize, learn, and reclaim urban space. A growing movement of stakeholders – citizens, activists, non-profits, and policy-makers – is interested in making sure that urban agriculture is officially recognized and supported by city policies.

The Office of Sustainability, along with other city agencies, saw the city's strong network of activists engaged in community gardening and urban agriculture[16] in Philadelphia as potentially an important vehicle for addressing the city's economic, environmental, and social challenges. City officials recognized the potential for "Farmadelphia," the notion that Philadelphia could build on its past agricultural history and bring farming into the city as a profitable endeavour that could stabilize communities, create jobs, and help meet local food needs.[17]

City officials have also been forced to respond to the demands for formal recognition of urban growing made by a group of younger urban growers. More recently, a coalition of social and racial justice activists have been demanding community land access and tenure and more formal participation in city governance. However, the city's response to demands for greater institutionalization of urban agriculture is not as seamless as some activists had hoped. We discuss this more later in the book (see chapter 4).

Figure 2. A vacant lot used to grow food in South Philadelphia.

The Vulnerabilities of Philadelphia's Urban Gardens

While Philadelphia has a long history of growing in the city, in to-day's excitement surrounding urban agriculture, that history is often brushed over. One interviewee warned: "The thing that people forget ... there's all this interest now in gardening in Philadelphia, and worldwide. People keep writing these stories about, 'Oh there's so many more farms'... But the truth is, we're way down from where we were 20 years ago. That story is getting really hidden behind this movement of urban agriculture, which is tragic" (Interview 3).

Urban agriculture in Philadelphia is *not new*. In fact, enthusiasm for gardening has surged during previous economic depressions and ebbed during economic booms (Lawson 2005; Bassett 1979). The first formal, citywide urban gardening effort dates back to the

Figure 3. Southwark–Queen Village Community Garden, established in 1976 and maintained by a local garden land trust.

1890s, when Philadelphia, along with many other cities in the United States, was experiencing an economic depression. Philadelphia borrowed the idea of urban agriculture as a solution from Detroit. The success of that city's program inspired citizens and officials in Philadelphia. Over time, urban gardens have served as an inexpensive solution to the vacant lot and poverty problems in Philadelphia, and this activity has been supported through government funding at federal and city levels, as well as by extensive involvement and leadership from non-profits, such as the Pennsylvania Horticultural Society (PHS).

We provide an in-depth analysis of the history of urban gardening in Philadelphia in chapter 2. But it is important to highlight here that even with funding, institutional support, and broad interest in urban gardening, garden sites have always been vulnerable to displacement and redevelopment (Lawson 2004). In recent years – and particularly during the housing boom of the 2000s – many garden sites around Philadelphia have been threatened by retracted support for gardens, the aging of the gardening population, and pro–economic development city land policy (Vitiello and Nairn 2009). These examples of displacement or threats of displacement highlight some of the challenges many urban growers and urban agriculture enthusiasts are facing today. During an economic bust, when vacant land is plentiful and there is limited demand for that land, they flourish. During an economic boom, when developers take interest, gardens and farms are threatened. Here, the dual role that city officials play in both supporting urban agriculture and promoting development is an important part of the story this book tells.

This book highlights how city officials (after much continued lobbying by activists) have started to embrace urban agriculture as a positive activity that can promote urban sustainability, particularly through the planting of gardens on vacant lots, thereby potentially alleviating food insecurity in underserved, low-income neighbourhoods. City officials talk about the importance of urban agriculture in documents like *Greenworks*; that said, the city has to balance urban agriculture with competing (and often more profitable) land uses. Some city officials view vacant lots converted into farms and gardens as ripe for development that would generate more taxes. City officials are clear that Philadelphia's future is not the same as Detroit's, where some have proposed that entire sections of the city not be redeveloped, contending that they may make more sense as farms.[18]

The folding of urban agriculture into existing urban development trends and policies is not easy and raises important questions about place-making and decision-making around urban land, particularly in disinvested neighbourhoods. Today in Philadelphia, tensions are growing: urban gardeners and farmers, some who have been growing for years and some who are younger and perhaps new to the city, are demanding more access to land just as the real estate market is picking up.

Urban agriculture can be tricky because in many ways it works against traditional urban financing and urban policy-making. In their 1987 book *Urban Fortunes*, Logan and Molotch talk about a unique feature of cities: the "use value" and "exchange value" of land. They explain how city leaders work to maximize the exchange value of a city's property, primarily in order to expand the property tax base (Logan and Molotch 1987). This sometimes pits city officials against residents, who may be more focused on protecting and improving the "use value" of cities. In the "growth machine" model that Logan and Molotch lay out, urban agriculture can be "tricky" for city officials who are hoping to capture more development (and the revenue it brings in), because in most cases urban agriculture privileges the "use value" of land, as seen in New York City, among other places (Logan and Molotch 1987; Schmelzkopf 2002). Urban growers and activists who pressure city officials for access to land for growing food and for community green space are not primarily interested in raising the value of land. Instead, they want to "use" the land for growing, community building, and community empowerment. As we are seeing more and more in Philadelphia, tensions develop when increasing land's "use value" through urban agriculture simultaneously increases its "exchange value." Because they increase "exchange value," the very same urban gardens and farms that helped stabilize the community come under threat because of the pressure to redevelop neighbourhoods (unless the farmers and gardeners have legal land title). One urban grower in Philadelphia saw what happened in New York City when urban gardens and farms were redeveloped as a lesson for Philadelphia: "That's a major tension of urban farming in general, tax dollars versus permanent land for farming. Farming is cool now because it increases the value of land around it, but that's exactly what happened in NYC, as soon as the land becomes valuable enough that there's a 'more valuable use for the land,' they're going to kick the farmers off, as this whole thing blows up in Philly" (Interview 15).

Proponents of urban agriculture in Philadelphia worry about the future of growing in Philadelphia, particularly as the city starts to see population growth. The experience in New York, where many community gardens have been redeveloped, highlights what happens to community spaces as the demand for development increases (Smith and Kurtz 2003). Activists who lobby for more urban agriculture worry that if promoting it makes the city more desirable for development, the same processes of rapid development and garden eradication that happened in New York City will happen in Philadelphia. They are working to develop mechanisms for more secure title to the land in order to protect Philadelphia's gardens and farms.

In many American cities, urban planners have long had an ambivalent relationship with urban agriculture (Lawson 2004; Thibert 2012). This is not because planners are necessarily opposed to urban agriculture. Instead, as Thibert (2012) notes, there are often cultural, technical, and legal barriers to implementing urban agriculture in the city. Lawson (2004) argues that gardens "receive praise as illustrations of local action to serve environmental, social, and personal needs" (151); however, such praise is often based on the notion that they will be temporary. When urban agriculture becomes more permanent, the tension between use value and exchange value becomes more problematic. As urban agriculture activists increasingly demand that urban agriculture become a permanent feature in cities, planners and policy-makers will have to figure out how to balance the demands by urban farmers and growers that gardening and farming spaces be promoted and protected with other critical community and economic development needs.

In Philadelphia, activists argue that urban agriculture is a legitimate and beneficial urban land use, and we see city officials responding to the demand for urban agriculture by changing policy documents and developing mechanisms for community engagement in land use decisions. City officials along with activists have been incorporating urban agriculture into the city's comprehensive plan and zoning codes, recognizing urban agriculture in the city's vacant lot redevelopment strategies, establishing a Land Bank to increase legal access to land for development and community spaces, and streamlining city-owned land disposition processes. They are also beginning to engage in food security planning at the city and regional scale.[19] But the process is uncharted and often uncertain, implementation is slow, and a politics of inclusion and exclusion has privileged certain voices – often professional, white, and politically connected – in the policy-making process, although this is changing. The story of Philadelphia's efforts to institutionalize urban agriculture within its policy framework is instructive for other cities because the political dynamics around urban agriculture in Philadelphia are constantly evolving. Philadelphia today is home to an interesting coalition of activists who are a part of the policy dialogue about growing food in the city. There is a strong network of activists who have been engaged with urban growing for decades; there is also a network of young gardening and farming activists who have been arguing that the city needs to institutionalize urban agriculture. These younger activists, many of them white, and some of them dismissed as "hipsters" or "millennials," are increasingly engaging with the city's non-profits, with older gardeners from different socio-economic and racial backgrounds, and with government agencies to make sure that urban agriculture stays on the city's redevelopment agenda. More recently, a group of younger farmers of colour, who see urban agriculture as a means to promote social justice and community empowerment, have joined the policy conversation through groups such as Soil Generation. They see urban agriculture as an important part of a larger, ongoing, multi-generational struggle for racial and economic justice, community nourishment, community control, and self-determination. They have been changing the conversation about urban agriculture and encouraging some of the younger white farmers and growers to explicitly address issues of race and class in their work. In Philadelphia today there is an expanding and vibrant discussion about the social and racial justice implications of having a group of activists, often young, white millennials disproportionately influence land use policy, which is primarily targeting land in disinvested, often African American communities. Increasingly, there are calls to meaningfully engage lower-income residents of colour in decision-making about food policy and urban land access in their neighbourhoods. Together, urban agriculture activists, young white and black farming activists, and older farmers and gardeners are working to highlight the ways in which urban agriculture can help meet social and racial justice goals as well as the ways in which urban agriculture can be problematic when it is tone deaf to the underlying race and class dynamics. These dynamics among different demographic groups, all with different relationships to power, are what make the Philadelphia story particularly interesting. The way the conversation about urban agriculture has been evolving in Philadelphia is instructive for other cities because it touches on issues of power, privilege, race, and class and raises questions about who decides what the future of land use in a sustainable city should look like.

Plan for the Book

This book traces the evolution of urban agriculture in policy and planning circles in Philadelphia as it has become an increasingly integral component of urban sustainability. Through interviews with urban growers and policy-makers, it explores how urban agriculture can be integrated into mainstream city land policy. It chronicles the shift in the discourse, acceptance, and long-term viability of growing in the city as city policy towards urban agriculture has evolved. We analyse the large-scale dynamics that have influenced the role of urban agriculture in Philadelphia over time.

We take a step back and look at how urban agriculture fits into the larger picture of urban redevelopment in Philadelphia. The book starts with a historical overview of urban agriculture in Philadelphia and then examines the evolution of urban agriculture there, primarily from 2008 to early 2017, a period during which city officials framed their work in the context of becoming "the Greenest City in America."[20] There is a strong movement of activists in Philadelphia who see promoting urban agriculture as vital to the city's long-term commitment to sustainability. Through interviews, we engaged with gardeners and farmers to understand their motivations and their perceptions. According to them, for a city to be sustainable, it must support urban agriculture as an important use of city land. They see urban agriculture as challenging the existing food system, providing access to healthy food, and offering vital community and environmental benefits. They also argue that access to community-controlled land is important, particularly in low-income, minority communities where gentrification is a growing threat. We see the story of Philadelphia's efforts to incorporate urban agriculture into urban policy as instructive for other cities where citizens, activists, and policy-makers are also trying to figure out the role that urban agriculture will play in making cities more sustainable. Some of the lessons from Philadelphia may be worth repeating; other lessons are more cautionary.

Chapter 2 describes the role – in many ways fleeting – of urban farming in city planning and policy in Philadelphia from the late 1800s to the present. Vacant lot cultivation in Philadelphia was born out of the economic depression of the 1890s. Arguably – and interestingly – the most recent interest in urban agriculture was fuelled in part by similar economic circumstances. By reviewing the history of urban agriculture as an alternative to charity, a tool for neighbourhood stabilization, and a sign of community organizing, this account situates and contextualizes urban farming and highlights some of the persistent challenges faced by the farming community, from waning interest in farming during economic boom periods to persistent land tenure issues and availability of resources. This chapter sets the stage for understanding what is new about the current interest in growing in the city and how urban agriculture is shaping the politics of sustainability in Philadelphia.

Chapter 3 traces the multiple paths pursued by various urban gardening programs in Philadelphia from 1990 to the early 2000s. The presence of gardening in urban planning, youth programs, and recreational activities reflects the different agendas that incorporated this activity, movement, and vision at multiple scales – regional, neighbourhood, individual – and many different levels of investment, including government, philanthropic, volunteer, and resident. Through an analysis of media, including newspapers, websites, and blogs, the chapter explores the role – or perceived role – of gardens as an approach to urban economic development, a form of social services for youth, and a guiding vision for a new Philadelphia. This analysis highlights how deeply embedded the gardening movement has become in a city in transition and, simultaneously, the role that transitional gardens can play in a city that is experiencing change, notwithstanding the historic and persistent reluctance to incorporate urban agriculture into the long-term vision for Philadelphia.

Chapter 4 examines how "urban greening" through "urban agriculture," "community gardening," and "green infrastructure" in the past several years has gone from being seen as peripheral to urban development to increasingly becoming a part of the urban sustainability agenda. City-led projects have highlighted the need for a more systematic citywide approach to urban agriculture. However, the tension between "use" and "exchange" value in Philadelphia continues to make it difficult to develop coherent policies around urban agriculture. The chapter highlights efforts in the city to formalize urban agriculture through the development of the Food Policy Advisory Council (FPAC), the inclusion of urban agriculture as an as-of-right use in the city's zoning code, and the development of the Land Bank and a more streamlined land acquisition process. However, the formalization of urban agriculture is a slow and complicated process with a lot of negotiation over city roles and responsibilities. Critics question how representative FPAC really is, and a growing movement of advocates has been pushing the city to address issues of social and racial justice associated with land tenure and community self-determination. The incorporation of urban agriculture into the city's revised zoning code and the recent development of Philly Landworks and the Land

Bank have formalized urban agriculture and provided a vehicle for streamlining access to vacant land. Urban agriculture is now written into the city's policy documents as an accepted land use with important value to communities. However, the terms of the Land Bank are still under development, as are those of other policies; a new *Land Bank Strategic Plan* was issued in 2017.[21] This all has taken place, though, in tandem with the rapid redevelopment of vacant lots in Philadelphia; some fear that the Land Bank may have been established too late to protect many urban gardens and farms. Hopefully, efforts to connect the Land Bank with the Neighborhood Gardens Trust (NGT), in combination with the Department of Parks and Recreation's Farm Philly's efforts to protect gardens and farms, may provide an answer for long-term garden preservation. Some urban growers worry that "these gardens are really hanging in the balance because there isn't much security and there isn't any real path to security" (Interview 36). In addition, with increased city interest in developing policies around urban agriculture, new questions have arisen about the safety of growing food in urban soils (and how to monitor it), access to water, the Philadelphia Water Department's stormwater fee, garden insurance, and the long-term sustainability of urban agriculture projects in general. The process of formalizing urban agriculture is iterative, and this chapter highlights the policy learning in the city around urban agriculture.

Chapter 5 tells the story of a "new generation of growers": individuals coming to Philadelphia or living in Philadelphia and joining or starting an urban farm or community garden. Urban farmers and gardeners recognized that Philadelphia's urban landscape represented both a challenge and an opportunity, and they took action. They saw Philadelphia's estimated 40,000 vacant lots – a number that sounds staggering to people from other cities – as an opportunity, even an asset. With enough sweat equity and sheer determination, vacant lots that were a symbol of blight and abandonment could be transformed into spaces of hope and inspiration. Healthy food could be grown. Communities could be reconnected with one another and with the land. This chapter explores the motivations of this group of primarily white, college-educated young people, who embraced urban agriculture as a way of life and who worked to turn urban agriculture into a viable economic development strategy. This movement is intricately connected to Philadelphia's long history of urban agriculture dating back more than a hundred years; however, the connections between the historical movement and the current one are often misunderstood. We juxtapose the more traditional farmers and gardeners in Philadelphia with this "new wave" of activists and highlight the need for more research into the demographic make-up of urban growers and their relationships to existing communities. This chapter also begins to uncover some of the complicated race and class dimensions of urban agriculture in Philadelphia, thereby raising questions about the relationship this new group of urban growers has with residents from disinvested communities.

Chapter 6 examines the role that urban agriculture in Philadelphia can and should play today and whether or not it is long-term sustainable. We examine cases where urban farmers and growers, who were committed to growing in the city, faced many challenges, from limited land tenure options to economic uncertainty. This chapter puts growing in the city in perspective by looking at the gap between expectations and reality and by asking questions about what the future holds for urban agriculture.

Chapter 7 concludes the book with a summary of key findings and with lessons for other cities. We highlight the ways in which the Philadelphia story raises important new questions for urban policy-making regarding how to connect urban agriculture and urban sustainability.

Methods

Our study of urban agriculture includes the following types of gardens, from McClintock's (2014) typology: allotment, guerrilla, collective, institutional, non-profit, and commercial. In short, we have included all urban garden types except residential gardens on private property that only served the residents of those units.

We aim to be inclusive in our language to reflect the diversity of the urban agriculture community, and we use many associated terms such as growing, community gardening, urban gardening, and urban farming.

The research for this book consists of archival research, media content analysis of newspapers and magazines, semi-structured interviews, and an analysis of current planning policies. The archival research was conducted at the Historical Society of Pennsylvania, the Pennsylvania Horticultural Society library, and the Temple University Urban Archives. The media content analysis of local newspapers includes a review of the *Philadelphia Inquirer* from 1970 to the present as well as *Grid Magazine*, a local sustainability publication produced in Philadelphia. Together the archival research and media content analysis provide historical accounts of urban agriculture in Philadelphia and shed light on the perceptions and representations of urban agriculture in official documentation as well as the positioning of urban agriculture in urban policy and planning by different organizations, authorities, and advocates. We also use current newspapers, websites, and government reports to highlight policy changes with regard to urban agriculture.

This work is supplemented by semi-structured interviews with more than thirty-five urban gardening stakeholders and enthusiasts in Philadelphia, conducted between March 2011 and December 2012 (and follow-up interviews with key stakeholders in 2013, 2014, 2016, and 2017), who helped us explore the state of urban agriculture in Philadelphia today. One of our primary goals was to understand what people talked about as a "new wave" of urban agriculture in the city. We were initially interested in the motivations of young, often (but not always) white farmers and how they understood the work they were doing, particularly in low-income, minority communities where they were outsiders. We think this is an important story to cover since the "new wave" of farmers and growers have been highly vocal about the need for policy-making to support growing in the city. In many ways, for better or worse, their view of urban agriculture and the role it plays has been shaping the policy conversations in Philadelphia. In our later rounds of interviews, we specifically reached out to African American urban agriculture activists to ensure that their voices were heard. We wanted to understand their critique of the "new wave" and to examine how their perspectives had been included in or excluded from more formal city policy-making around urban agriculture and land use. While we highlight how race and class is included and excluded from recent urban policy-making around urban agriculture in Philadelphia, we fully recognize that a more critical analysis of the racial politics of urban agriculture is the topic for another, much needed book.[22]

During the interviews, we asked the participants why they had involved themselves in gardening in Philadelphia. We also asked for their impressions of their work's impact on the community, neighbourhood safety, food security, and property values, and about whether they had seen changes in government or other outside support for urban agriculture. Our understanding of the politics of urban agriculture in Philadelphia also comes from our own involvement in the planning process. Both of us served on the FPAC Subcommittee on Vacant Land (now Urban Agriculture) and participated in FPAC's Soil Safety Working Group. Several of our students have gone on to start urban farms or work for the city or for non-profits, and many of them have been active in urban agriculture. The findings of the book are our own and do not represent any official position of any organization.

Contributions

Popular and scholarly interest in urban agriculture is growing. This book provides a critical perspective on the development of urban agriculture policy within the broader urban-economic context of a large city struggling to manage competing sustainability objectives. Previous urban agriculture research has focused largely on the community dimensions of urban agriculture and the potential for this activity to make a positive difference for neighbourhoods. In contrast, our research marks one of the few efforts to reflect on the role of urban agriculture in the urban policy-making context and on the underlying tensions that are often oversimplified or obscured by enthusiasm for urban gardens and farms. We start from a historical perspective, track the current political developments around urban agriculture, listen to stories about how the "new wave" of urban growers make sense of their work, and then connect perceptions of the role of urban agriculture with the reality of what is happening in Philadelphia. This book strives to contextualize urban agriculture among the many other economic, political, and social dynamics in a post-industrial city seeking to transition to a sustainable future. Our goal is to begin to tell a more comprehensive story about urban agriculture in Philadelphia and to pull out lessons that can be applied to other cities.

We begin with a historical overview in chapter 2.

chapter two

The Rise of Urban Agriculture as a Short-Term Strategy (1893–1980)

While there is a lot of discussion about the "green wave" of urban sustainability in Philadelphia, a wave that includes urban agriculture, it is important to ask: What is new about the current enthusiasm for urban agriculture? Philadelphia has a history of urban growing, which has been supported by a multitude of institutions since the late 1800s (Figure 4). Gardens have long been a part of Philadelphia's landscape, community, and politics and planning. That said, many gardening and sustainability enthusiasts argue that the current trend in gardening represents a moment or opportunity for something different from before. This chapter reviews the history of urban gardening and farming in Philadelphia since 1890 and asks what, if anything, is new about the current gardening movement. It focuses on how urban agriculture changed from the late 1800s to the 1970s with respect to its objectives and agendas, purported benefits, key audiences, and policy practices. It shows how urban gardening evolved from a food production strategy for economically vulnerable individuals led by a centralized organization to a more diffuse grassroots effort, supported by numerous institutions, to beautify and build community in neighbourhoods experiencing abandonment and blight. This account also reveals the tangential ways in which urban agriculture was incorporated into city land policy as well as the long-term implications of this tenuous relationship for growers and their gardens.

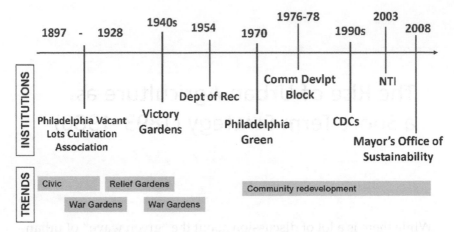

Figure 4. History of urban gardening and farming in Philadelphia.

For decades, gardens in Philadelphia have served to manage vacant spaces, organize communities, and provide food, however vulnerable they are to development pressures, land scarcity, and limited long-term acceptance in policy circles. Tracing these vulnerabilities over time allows us to consider what is unique about the current trend in gardening and whether the sustainability movement has the potential to better sustain gardens over time.

Early Gardening Efforts from 1890 to 1930: Vacant Lot Cultivation

As you might have already heard, the corner of 20th and Market has been transformed. About a month ago the space was a fenced-in vacancy, covered in weeds and trash, a more than 20-year-old eyesore in Philadelphia's business district. But now, amidst the skyscrapers of Center City, stands what Pennsylvania Horticultural Society (PHS) President Drew Becher calls a "billboard" for urban gardening in Philadelphia.[1]

The quote above, published in a popular Philadelphia magazine in June 2011, implies that the conversion of vacant lots to gardens and farms is a new and innovative approach to urban land reclamation during tough economic times. However, as documented by the research of Bassett (1979) and Lawson (2005), the vacant lot cultivation movement in the United States was born out of the 1893–4 economic depression. That economic downturn led to high unemployment rates across the nation because of widespread bankruptcies among businesses and railway companies. Lawson (2005) writes that in 1894, the mayor of Detroit, Hazen Pingree, suggested that unemployed labourers be employed in a vacant lot cultivation project. The labourers would be provided with land by willing landowners, and with financial support and technical guidance from the city, and with these, they could grow produce for their own consumption and for sale. The program would provide the unemployed with work and put vacant land into productive use.

Like other Progressive Era practices, the vacant lots program was described in newspapers as a favourable "alternative to charity" during a time of crisis.[2] Progressives saw charity as a superficial solution that failed to address the underlying causes of poverty; in their view, simply giving money or provisions to the poor only reproduced the cycle of poverty and "pauperized" families. Gardening provided an alternative. The vacant lots cultivation program created opportunities for the unemployed to grow their own food and thereby become more self-reliant. Proponents of that program also highlighted its moral and ethical dimensions. Instead of becoming preoccupied by financial stress and the lack of work, the unemployed would be engaging in a productive and rewarding activity that would sustain their families during challenging times. The language employed by the Philadelphia Vacant Lots Cultivation Association in their marketing materials reflected these ethical and economic objectives: "While contributing to their health and happiness, and receiving valuable training and experience, the men, women and children join in increasing their material supplies. As their own work produces the results, they are not pauperized, but encouraged to be more industrious and self-dependent, and acquire greater ability and self-respect."[3]

Furthermore, and as Bassett (1979) described, gardening would serve as a social institution for the most vulnerable populations during the economic depression. Supporters of gardening programs saw them as a short-term response to an emergency – indeed, as an ideal solution to unemployment, because those programs would provide work, food, and income. Unemployed labourers could cultivate the land until the economy rebounded, and with it employment opportunities.

Lawson (2005) writes that even though it dovetailed with other Progressive Era philanthropy programs of the time that emphasized self-help over charity, the "Detroit Experiment" initially met resistance and scepticism among churches, philanthropic organizations, and landowners. The effort was highly politicized and was initially funded by Mayor Pingree out of his own pocket. But after the estimated value of the crop yield far exceeded the initial investment, the program was deemed successful enough that its continued existence could be justified. On the heels of that success in Detroit, other cities experiencing similar economic hardships formed their own vacant lot associations.

The Philadelphia Vacant Lots Cultivation Association (PVLCA) was launched on 6 March 1897. The city's mayor, Charles Warwick, was appointed Honorary President of the Advisory Board.[4] In its first year, the PVLCA under Superintendent Powell supervised twenty-five acres, mostly in West Philadelphia (Lawson 2004). This land was divided into gardens for one hundred families, each paying fifty cents for tools. The media deemed the first year a success, and this paved the way for the program's expansion the following year.[5] The local paper reported that Powell was enthusiastic about the program; indeed, he predicted that it could easily support 2,000 and possibly up to 5,000 families.[6]

The PVLCA assigned plots of one-sixth of an acre to each household enrolled in the program and provided seeds, fertilizer, and technical gardening assistance.[7] Each plot cost the PVLCA $5 annually, and by 1904 gardeners were required to pay $1 per garden for seed to address the PVLCA deficit.[8] The gardening program grew in popularity, and applications for plots increased annually. By 1900 the superintendent could record in the PVLCA minutes that there were 492 lots under cultivation; he requested additional land to meet the rising number of applications.[9] Two years later he reported that 700 gardens on 157 acres were under cultivation.[10] In addition, the quantity and quality of the yield had improved, factoring in weather conditions.[11]

The PVLCA's popularity grew. The local newspaper reported that besides combating high unemployment rates in the 1890s, gardening helped reduce the cost of living during and after the First World War, particularly for families that had suffered injury or loss.[12] As evidence that the program served more than the lowest-income families, a summary from the 1921 growing season indicates that the 664 gardeners were from a wide range of occupations: labourers (237), factory hands (106), building trades workers (86), railway workers (50), clerks (47), tradesmen (45), municipal employees (24), teamsters and chauffeurs (22), widows and old men (14), and miscellaneous (33).[13] The local paper applauded this broad service, noting that garden plots were often assigned to populations designated as vulnerable, such as those who worked in "confined settings" (which were believed to be poor for one's health), the elderly, and widows, although the gardeners were of all ages and nationalities.[14] The program, which lasted for more than thirty years, remained relevant through many historical events, from economic depression, to times of inflation, to wartime. One reporter noted in 1917 that "everybody would like to be a farmer just now, with several hundreds of barrels of white potatoes, sweet potatoes, onions or turnips stowed safely away, to sell off at top-notch prices."[15]

Philanthropists and government institutions deemed the program a success, particularly for economic and ethical reasons. The *Philadelphia Evening Bulletin* reported that vacant land cultivation was more than a "feel good" charity for donors – it generated high returns from donations. By the 1920s the financial value of crop yields ranged between $55,000 and $100,000.[16] Gardeners often produced enough to feed their families and sell the surplus. PVLCA superintendent Dix explained in an article published in the local paper the economic opportunities afforded by the program, for gardeners and donors:

> We can find lying idle at the present time, 1,700 acres of good land in the city ... This would supply gardens to 10,000 families and cost no more than $50,000. If $50,000 were set aside for the relief of 10,000 families, it could not provide more than $5 worth of food per families. But $50,000 applied to starting families on gardens could result in at least $500,000 worth of food or an average of $50 per family.[17]

Proponents of the program also emphasized the long-term benefits for the gardeners. From the start, the program was trumpeted as an alternative to "pauperization" and idleness. And given that most gardeners enrolled in the vacant lots program were from urban settings, proponents and philanthropists hoped that the agricultural training they received would lure them into farmwork or at a minimum inspire them to maintain their own gardens. Dix reported on the benefits of gardening in his 1912 annual report to the PVLCA:

> The value of manual training today is generally conceded, and there is no line of manual training which has such a broad influence as gardening. No other training stimulates as much imagination, originality and initiative, and no other training builds up as much practical thought, foresight, and feeling of personal responsibility as ours.[18]

Interestingly, Lawson's (2005) research suggests that farming was viewed as a promising occupation at this time, one that held the potential for specialization and that entailed few start-up costs. The program became increasingly popular because it provided benefits to the city and to landowners, as well as to the gardeners themselves. The cultivation of vacant lots beautified idle land that one news reporter writing in the 1920s characterized as "dumps" and "an eye-sore to the observer."[19] In an article he published in the local paper, superintendent John K. Snyder touted the "economic, esthetic, and sanitary" features of the program[20] and encouraged landowners to contribute their vacant land to the program instead of hoarding it as an investment. He often noted that landlords were pleased with the gardening activities being carried out on their lots; according to him, few landlords complained.

Despite the initial successes, vacant lot cultivation raised some thorny issues among the parties involved (Lawson 2005). When ethical objectives collided with economic ones, there was discussion about transferring the overhead costs to the gardeners by asking them to pay for the land they were cultivating. A graduated fee was implemented, with first-time gardeners paying one dollar per plot, second-year gardeners two dollars, and so on.[21] These fees did not always cover the costs of ploughing, purchasing seeds, or fertilizing the soil, and the PVLCA was required to take out loans to address the deficit.

Consistent land availability was a critical issue throughout the program, according to the PVLCA.[22] Another was land tenure, which was debated among officials and landowners, and which points to the short-term perspective on urban agriculture in urban land policy and practice. Landowners resisted the idea of making their land freely available to gardeners in perpetuity. Some of them asked that the gardeners pay rent; others were reluctant to rent land, choosing instead to hold it for "speculation."[23] Regardless of the specific land tenure arrangement, the gardens were largely viewed as temporary. Lawson (2004) examined garden locations from 1899 to 1925 and found that few vacant lots remained gardens during this time. In many ways, this temporary status for gardens was vital to the program's initial success, in that landowners would have been far more reluctant to donate their land to the gardeners either permanently or for a long period of time. Donated land was subject to return to the owner on ten days' notice, and landowners were not liable for crop loss, according to the superintendent.[24] Even under policies that favoured them, landowners often hesitated to lease their land to PVLCA. The local paper reported that one landowner refused to provide land out of fear that the gardeners would begin to think they owned it: "You let these people cultivate the ground and after a while they will get to thinking they have a natural right there."[25] The PVLCA maintained that this fear was not general;

nevertheless, it struggled to obtain enough land to meet the growing number of applicants each year. Superintendent Dix pleaded each year for more landowners to consider the vacant land program: "I want to ask our directors, members and friends, to bring all the influence they can to bear upon such owners, in order that they may see our work in its true light and realize what a great benefit they can be to the poorer workers of our city, by simply loaning the temporary use of their idle land without depriving themselves of anything whatsoever."[26]

Because growing food was encouraged during the war, the program peaked in 1918–19. Then the postwar period led to increased employment opportunities. Throughout the program's existence, land tenure remained a critical issue, especially after the First World War. In November 1918 the building industry was released from government restrictions "as a means of facilitating the general industry of the country from a war to a peace basis."[27] The government saw this end to restrictions as a way to meet domestic industry's needs and to boost employment. With the local paper boasting of construction as the next "key" to revitalizing the economy,[28] a building boom ensued, with factories leading the trend. Within a year, the number of construction projects exploded in Philadelphia, summarized succinctly by the following headline: "West Market St. Undergoing Boom: Thoroughfare between fifteenth and Schuylkill Changes Vastly in Decade. Increase in Realty Values Indicates how Street is Growing."[29] This article reported that property values on Market Street had doubled, tripled, and quadrupled in a decade and that difficulties "financing construction work have proven no bar to [the] construction boom." As a result, the demand for vacant land rose steadily. In 1924, garden superintendent John K. Snyder reported to the local paper that he had started looking for cultivable land along the periphery of the developed areas.[30] This displacement pattern is consistent with the idea that gardening had been incorporated into urban land policy as a temporary land use rather than as the highest or best use of the land. As development picked up, it was privileged over urban agriculture.

By the 1920s the economic upturn was having a direct impact on the vacant land cultivation program. Lawson's (2005) research indicates that in 1923, 197 plots were lost due to the economic boom and associated development and construction (Lawson 2005, 39). These gardens were not replaced. In March 1924, superintendent Snyder was quoted in the local paper: "Philadelphia's building boom is making it hard to carry on the work of turning vacant lots into vegetable gardens for thrifty people ... Many of our old plots, where hundreds of gardens were planted, are now sites of rows of houses."[31] By 1927 the demand for garden sites was low and available land was scarce. In a booming economy, there was little space for vacant lot cultivation.

On 10 November 1927 the president of the PVLCA, Samuel Fels, called a special meeting to discuss whether the association should continue or disband. Superintendent Snyder pointed out several challenges the program was facing, including vandalism, improved employment conditions, the decrease in available land, Japanese beetles, and new construction. The PVLCA decided to continue the program for one more year, to follow through on commitments to "several hundred families."[32] In September 1928 the PVLCA decided to complete its existing obligations and then to "cease to exist," primarily due to the lack of available land and the falling demand for gardens.[33]

The vacant lots cultivation program ended in Philadelphia in 1928 because of limited land and weak demand for garden space;[34] however, the gardening movement grew popular again during the Second World War in the form of Victory Gardens. The National War Garden Commission's program aimed to grow food for domestic consumption so that commercial produce would be available for the war effort overseas. Bentley (1998) observed that in the 1940s, planting Victory Gardens was an act of patriotism rather than a way to ensure individual or family survival as in earlier eras. Because we focus in this book on "relief gardens" or those designed to provide families and individuals with resources, as opposed to war gardens promoted as part of a national effort, this book does not provide an in-depth assessment of Victory

Gardens. Many other accounts describe such gardens (e.g., Miller 2003; Bentley 1998; Kurtz 2001).

"Flower Power" in the 1970s

Hynes (1996) notes that in the postwar era, a transition took place across the United States: gardening turned away from being a primarily civic duty associated with supporting a nation at war towards being a social movement focused on empowering communities affected by economic hardship and abandonment. In some ways this transition echoed the early days of urban agriculture in Philadelphia, when it was a stopgap measure in times of economic hardship. But rather than following a top-down approach coordinated by a central organization such as the PVLCA, in the 1950s community gardens became a form of grassroots activism in response to social and economic upheaval in many American cities (Kurtz 2001). The trend towards deindustrialization and suburbanization across North America led to population decline in many cities, including Philadelphia. In 1960, Philadelphia had 2,071,605 people; by 1990, it had had lost one-quarter of its population (US Census).

Population decline led to increases in vacancy rates and in the number of abandoned buildings. By the 1970s, various organizations, institutions, and residents were turning to community gardening as a way to reclaim their neighbourhoods and to address food security concerns in the face of population loss, abandonment, and urban renewal (Lawson 2005). Lawson (2005) links these changes in community gardening to the evolving role of gardeners from the 1950s to the 1970s. In early years, a community garden typically referred to a garden site divided into individual plots for different gardeners who tended to their own respective plots. Also, the purpose of community gardens in the early era was twofold: educational and social. Community gardens were educational in the sense that they served as demonstration gardens to teach local people best practices of planting, maintaining, and harvesting for the purpose of increasing food production. But these gardens were also viewed as an ideal way to bring people together at the same garden site, even if they were working in separate beds.

By the 1970s the concept of community gardens had expanded to include community involvement for the purpose of community empowerment. Lawson (2005) notes that community gardening in the 1970s was marked by more gardener involvement in planning and developing projects. In previous eras, philanthropic or civic groups prepared the garden sites; gardeners then applied to use those sites. By the 1950s more gardeners were actively preparing sites, negotiating land tenure with city officials, and mobilizing support to oppose the redevelopment of gardens. Kurtz (2001) highlights the multiple uses and layers of meaning associated with gardening. While matters of food production and food security continued to steer community garden projects, some gardeners were also beginning to think of their gardens as spaces for reflection, community gathering, or land reclamation, in addition to the cultivation of produce.

Proponents of community gardening contended that this collective work would improve food security, strengthen social networks, and build local capacity; all of these benefits could then be extended into other neighbourhood concerns, such as reducing crime or cleaning up streets.[35] This theme of self-management and self-reliance is interesting to consider in the institutional context of community gardens. Even while the gardens focused on community, multiple institutions and organizations supported and enabled these projects during this transitional period. This support suggests that to some degree, the city accepted the validity of urban agriculture. Furthermore, given the extent of abandonment and vacancy in some neighbourhoods, questions of land tenure were less pressing than in previous eras, at least until economic redevelopment rebounded and development proposals threatened the gardens.

Community-Run but Institutionally Supported gardens

Community gardening in Philadelphia owes much to ongoing institutional support (Figure 4). Government grants as well as non-profits such as the Neighborhood Gardens Association (NGA) and the Pennsylvania Horticultural Society (PHS) have been instrumental in supporting community gardens since the 1950s. This institutional support has often relied on city, state, or federal sources for resources such as planting materials and technical support and training and indicates that urban agriculture has long had one foot in formal city land policy, even if the foothold has been tenuous, with objectives shifting over time.

Increasingly, the objective of institutional support programs has been to build community capacity through urban greening and gardening, rather than to produce food as in the early 1900s. All of these programs acknowledge the community's vital role in sustaining gardens over time. They have often provided start-up funds and technical support for gardens, but community support and labour have been necessary to keep the projects going. Three of the most influential institutions and their changing objectives over time are described briefly below to contextualize their work as the gardening movement has placed more emphasis on community.

Neighborhood Gardens Association

In Philadelphia, the transition to community gardening in the postwar era is often traced back to the efforts of Louise Bush-Brown, director of the Ambler School of Horticulture for Women until 1952, who founded the Neighborhood Gardens Association (NGA) in 1953 (Hynes 1996). The association was formed after Bush-Brown met with several other community leaders, including the directors of several settlement houses and community centres, the administrative head and the home adviser of a large public housing project, the principal of a large elementary school, and a number of citizens concerned about urban blight. They voiced their concerns about the growing number of underprivileged neighbourhoods across the city and discussed approaches to revitalizing neighbourhoods. Bush-Brown described the conclusion of their initial meeting: "It was the consensus of feeling on the part of those attending the meeting that if a few constructive garden projects could be initiated they might prove to be a catalyst which would stimulate an upsurge of community pride" (Bush-Brown 1969, 8). Clearly, the newly formed garden association was intended to provide more than beautification.

The NGA's mission was to foster interest in flowers and gardening in the city and to create civic pride around garden projects. Its official motto: "Strive to make the world a little better and more beautiful because you have lived in it" (Bush-Brown 1969, 8). The NGA's early horticultural initiatives involved planting window boxes and other small gardens in low-income neighbourhoods with little green vegetation. While the short-term goal was beautification, these plant boxes were also intended to spur community development. During the first year, 12,000 plants were planted, 427 window boxes were installed, and 800 dooryard projects were created in housing projects across Philadelphia (Bush-Brown 1969).

The NGA quickly set out to achieve its horticultural ambitions by connecting suburban garden clubs with urban neighbourhood groups. The association promoted long-term sustainability through the establishment of garden blocks in urban neighbourhoods. Any low-income block could apply to become a garden block, the key requirement being that 85 per cent of the block agreed to maintain the garden and that one person would serve as a leader and liaison between the garden block and the NGA (Bush-Brown 1969). Garden blocks were subsequently connected with a garden club affiliated with the NGA, which would sponsor the garden block for two years. Women from the garden club would bring flowers and planting materials in the hope that the initial planting effort would have a ripple effect across the neighbourhood (Hynes 1996, 107).

The NGA's initial efforts proved quite successful, and the demand for window boxes and garden blocks grew. By 1963 the NGA had helped establish 360 garden blocks, created 16 gardens on vacant lots, planted 1,200 dooryard gardens in housing projects, and developed 289 4-H club gardens.[36] More than acts of beautification, the window boxes became signs of leadership: "During the years since we initiated the program," said Louise Bush-Brown, "we have learned that it has developed leadership in areas where there's never been any opportunity for leadership."[37]

The NGA's community-based objectives were widely admired. In media reports about the NGA, words like "sense of pride," "neighborhood friendship'" and "hope" were associated with garden blocks and window boxes.[38] The NGA's work was recognized beyond Philadelphia, and in 1965, First Lady Mrs Lyndon B. Johnson invited Bush-Brown to the White House as a guest of honour.[39] Mrs Bush-Brown died in 1971; in 1977, the NGA, as originally conceived, disbanded due to lack of leadership.

US Department of Agriculture

The year the NGA disbanded, the US Department of Agriculture (USDA) launched an urban gardening program that provided funding of $1.5 million for urban gardening projects in New York, Chicago, Los Angeles, Philadelphia, Detroit, and Houston. These six cities were selected based on population size and number of low-income residents. Each state could have only one city as a recipient. By 1978, ten more cities had been added to the program. By 1989 there were twenty-three cities in the program, but funding increases were limited – $3.5 million had been allocated to the program. In cities like Philadelphia, where gardening programs already existed, the USDA extension provided technical and educational support. In Philadelphia, the Penn State Urban Garden Program developed demonstration gardens to teach gardeners best practices (Lawson 2005).

Echoing the goals of the 1900s-era gardening efforts, the program's initial purpose was nutritional – to help low-income residents grow produce and improve their health. As one proponent of the program noted, "gardening is one way people can help themselves, and so should be popular in the face of budget cuts in social welfare programs" (Blake 1981). The program was evaluated based on return on investment, measured by produce (Hynes 1996). The returns were respectably high for the program, with each federal dollar yielding $6 of produce in 1985. Hynes (1996) observed that the USDA Urban Gardening Program was extremely cost effective – it produced $22.8 million worth of produce on a budget of $3.5 million in 1989 – and despite a constrained budget, it succeeded in increasing the number of cities served, volunteers trained, acres cultivated, and the quantity of produce yield.

Once the program was deemed a success in food production, some advocates pointed to its potential for promoting urban and community development. Hynes (1996) lauded the program for its potential to provide the low-income families with self-determination, independence, and a land ethic. She cites the interviews conducted with 178 gardeners in Newark by an Essex County agricultural agent, Dr Ishwarbhai Patel, who found that gardeners benefited from healthy food, money saved, and a sense of community and friendship. Besides serving individuals and sub-populations, the gardens could become important anchors for community development.

"Gardening is really the medium … But we want to … empower the people in the community to have a better life, to learn responsibility, to learn how to cooperate." The preceding quote, from the director of the Pennsylvania State Urban Gardening Club in 1996, indicates that gardening was more than an approach to improving nutrition: it was also a mechanism for promoting social change (Gordon 1996). This motivation remains in line with the early vacant lots program from the 1890s, which points to an interesting yet widely unacknowledged consistency over many decades.

Philadelphia Green

Ernesta Ballard, the celebrated activist who turned the low-key Pennsylvania Horticultural Society (PHS) into a civic force after she took over its leadership in the 1960s, once said that "horticulture is universal."[40] The PHS has played an important and ongoing role in urban greening efforts around Philadelphia. It was founded in 1826 with funds from its annual flower show, which continues today as a popular annual event that attracts visitors from around the world. However, in the 1970s the PHS adopted a civic-inspired approach towards gardening. Ballard, the executive director of the PHS from 1963 to 1981, was inspired by Louise Bush-Brown's NGA program and felt that the members of an elite civic organization like the PHS should reach out to low-income Philadelphia residents for the purpose of community building. In 1974 the group launched some community outreach efforts that included working with schools, establishing community gardens and a junior flower show, and making efforts to landscape prisons (Bonham, Spilka, and Rastorfer 2002). When the city's funding for park programs and community gardens dried up in the 1970s, Ballard expanded the PHS's community outreach with the assistance of Blaine Bonham, who joined the group in 1974 and remained a leading figure in Philadelphia Green until his retirement in 2010. The community outreach program launched the 10,000 Trees program to add trees to city parks,[41] as well as the Gardenmobile program, which provided technical gardening assistance to community gardens in the city.[42] By 1978 the PHS had expanded this program in response to demand from other neighbourhoods. That same year it consolidated these many community-based efforts under one program called Philadelphia Green. This new program was launched with a grant of $250,000 from Mayor Rizzo's Office of Housing and Community Development, supplemented with $100,000 from the PHS, under the direction of Blaine Bonham and a staff of nine to work with community groups.[43] The types of community-based greening programs initiated under Philadelphia Green continue today and have become models for urban greening nationwide.

For Ballard, the philosophy of Philadelphia Green was similar to Bush-Brown's in that self-help and community development were key. But Ballard distinguished Philadelphia Green from other community outreach programs in that she sought mutuality, besides self-help. From its inception, Ballard viewed Philadelphia Green as a way to build bridges among people from different backgrounds, notwithstanding stark patterns of residential segregation along race and class lines. Ballard attempted to bridge these populations, not only by asking wealthy residents to fund Philadelphia Green but also through events that brought people together, such as city garden contests and harvest shows. The following quote from Gretchen Reilly, a judge at the 1994 garden contest, reflects Ballard's goals (Sokolove 1994): "This is something that has given me some of my fondest memories over the years. We have sat down with people, shared lemonade, gardening tips and expertise. It is just a wonderful experience." Reilly, a resident of Haverford, a wealthy Philadelphia suburb, was part of the team of four judges that year. Ballard hoped that volunteers as well as gardeners would benefit from Philadelphia Green.

The Philadelphia Green program was successful from its inception and quickly expanded its programs, staff, volunteers, and gardens. By the 1980s, community groups were cultivating 300 vegetable gardens, 200 garden blocks, 100 street- and curbside-planter blocks, and 58 sitting gardens (Lawson 2005, 255). By the mid-1990s the program had a staff of forty working with 1,100 neighbourhood groups, with 2,000 greening programs. Blaine Bonham, the financial manager of Philadelphia Green at the time, noted that the gardening program started out simply providing soil, planting materials, and technical advice. But as the program grew, its options for neighbourhoods expanded to include picnic areas, barbecue pits, sitting gardens, and so on. The program also sought to fully integrate gardens into their communities (Bonham, Spilka, and Rastorfer 2002).

The community-based aspect of Philadelphia Green was not simply written into its mission statement; it was also institutionalized in the gardening program protocols. The organization provided the initial financial resources and technical expertise to start a garden, but the long-term maintenance of that garden was the community's explicit responsibility. According to Bonham, "the aim was to create sustainable greening projects that would require minimal ongoing support from Philadelphia Green once the initial landscaping was complete" (Bonham, Spilka, and Rastorfer 2002, 76). To ensure the success of these gardens beyond the involvement of the PHS, the group required that 80 per cent of the block be committed to maintaining the garden space. This strategy ensured that the PHS only entered neighbourhoods with the residents' permission, support, and investment in the project over the long term.

Philadelphia Green expanded its scope from 1980 to 1993 through a program called the Greene Countrie Towne (Bonham, Spilka, and Rastorfer 2002). This program's name invoked William Penn's plan to develop a green city with substantial open space, as indicated in the following set of instructions written by him in 1681: "Let every house be placed, if the person pleases, in the middle of its plat, as to the breadthway of it, so that there may be ground on each side for gardens or orchards, or fields, that it may be a greene country towne, which will never be burnt and always wholesome."[44]

Bonham viewed the program as a "tribute" to Penn's original vision – indeed, one reporter described the effort as the "re-greening" of Philadelphia (Darling 1983). This greening strategy was also seen as a means to another end. The PHS Greene Countrie Towne program was designed to incorporate greening as part of the development process, with the ultimate goal of building the organizational capacity of communities. This emphasis on community development resonated with other organizations and funders. Bonham summarized the vision of Greene Countrie Towne: "This considerable shift in thinking and practice – a shift that two Philadelphia-based foundations were eager to support with significant ongoing funding – also signaled Philadelphia Green's growing ability to act as an agent of urban revitalization" (Bonham, Spilka, and Rastorfer 2002, 76).

Bonham, Spilka, and Rastorfer (2002) describe how the program collaborated with community-based organizations over three to five years with the goal of beautifying neighbourhoods and empowering residents. To ensure the sustainability of the greening effort, communities were required to demonstrate their support for the program. The criteria for inclusion in Green Countrie Towne included the following: median family income per block had to be at or below the city's median income; at least 60 per cent of the housing had to be owner occupied; residents on the block had to demonstrate a commitment to gardening; and there had to already be neighbourhood organizations. The PHS saw community buy-in as vital to a garden's long-term success. Eight "Townes" were selected and received a three- to five-year commitment from the PHS. This commitment included services such as sealing walls near vacant lots, removing debris, issuing permits to use hydrants for watering, and providing composting through the Fairmount Park Commission. PHS staff planted street trees, cultivated vacant lots, and created sitting gardens. They also promoted greening through educational programs and other events.

Ballard viewed the gardens as bridges between populations of different races and socio-economic backgrounds. But despite Ballard's insistence that the program be based on *mutual* involvement and respect, in the mid-1990s, Hynes (1996) noted an asymmetry in the programs. For instance, for the gardening contests, wealthy suburban gardeners were designated as judges for the inner-city gardens, without the reciprocal (i.e., inner city gardeners judging wealthy suburban gardens).

Hynes (1996) highlights the importance of the collaboration between the Penn State Urban Gardening program and Philadelphia Green. The directors of these programs provided the leadership, financial support, and direction needed to sustain and expand community gardening in Philadelphia through the 1980s, and in addition, the programs worked together. For instance, the Penn State Urban Gardening Program was not eligible for community development block grants from the city's Office of Housing and Community Development, so Philadelphia Green applied for the funds. The programs also collaborated on joint projects: the Penn State program spent funds on gardening and horticulture education; Philadelphia Green supplied physical and plant materials.

The two programs have also worked together to create an official mechanism for sustaining gardens over the long term. While the programs have successfully provided resources and training to cover the start-up costs associated with establishing a garden, neither institution can ensure the perpetuity of gardens. Both institutions assume that community commitment to the gardens will sustain them over time, but even community investment in the space cannot prevent certain types of encroachment, such as development.

To address a growing concern about the displacement/elimination of gardens, in 1986 the PHS and the Penn State program established a citywide urban land trust, named after Bush-Brown's program, the Neighborhood Gardens Association (NGA).[45] The purpose of the land trust is to ensure the permanence of the urban gardens. The establishment of the land trust helps address the question of land tenure for community gardens, which will be further discussed later in the book (see chapter 4).

While gardening has had a persistent presence in Philadelphia's built environment for more than one hundred years, the locations and types of garden projects have changed (Lawson 2004). Some of these changes have created conflicts over the future of gardening in the city. In addition to the shifting locations of gardens, the ambitions and interests in growing have also changed through time, raising important questions about the objectives and potential of gardening projects. These interests have ranged from economic development to neighbourhood stabilization, from beautification to food security, and from hobby to career. These same interests have led to initiatives supported by non-profits and city agencies. The following chapter explores the evolving role of urban agriculture in urban redevelopment initiatives, highlighting the ways in which urban agriculture serves as a proxy for a variety of social change initiatives.

chapter three

The Agendas of Urban Agriculture at Its Peak (1990–2000s)

This chapter considers the multiple agendas that featured urban agriculture during its peak popularity, from the 1990s to the early 2000s. The sheer number of gardens, estimated at around five hundred (Vitiello and Nairn 2009), during this era was the result of a city with abundant vacant land, a city government committed to greening as a beautification strategy, a host of non-profit organizations committed, financially and otherwise, to gardening, and a pervasive green identity among a growing number of residents. The presence of urban agriculture in urban planning, youth programs, and recreational activities reflects the incorporation of this activity, movement, and vision at multiple scales – regional, neighbourhood, individual – and many different levels of investment, including government, philanthropic, volunteer, and resident. This chapter explores the role – or perceived role – of gardens and farms as an approach to urban economic development, a form of social services for underprivileged youth, and a guiding vision for a future Philadelphia. This analysis highlights how urban agriculture became integral to many planning-related objectives in Philadelphia despite the insistence of policy-makers that the tenure of many garden and farm sites was temporary.

Gardening as an Approach to Urban Economic Development through Vacant Lot Reclamation

By the 1990s, depopulation had resulted in a growing number of vacant lots that created blight and attracted dumping and crime.

Philadelphia had close to 15,000 vacant lots and 27,000 vacant structures, each costing the city $3,000 annually to maintain (Hynes 1996). Concern over this situation led to a government initiative to reclaim vacant lots. The city's budget, and investment prospects, did not allow for the restoration of dilapidated buildings or for new construction on vacant lots; the city administration viewed plans for open space maintenance as a temporary solution to blighted land. At this time, gardening and vacant lot cultivation became more than a self-help initiative for communities; it was a city-endorsed strategy for urban redevelopment. While the benefits were to accrue to local communities (envision a maintained garden versus an overgrown and trash-strewn lot), the city also saw vacant lot beautification as a low-cost maintenance strategy until private developers showed interest in the vacant lots.

Community development corporations (CDCs) received encouragement and funding to incorporate open space planning into housing developments and to investigate options for vacant lots. Philadelphia CDCs sought the expertise of Philadelphia Green, the PHS's greening program, and this led to a close collaboration on vacant lot reclamation and beautification. This CDC–PHS collaboration opened the door to a direct partnership with the City of Philadelphia on vacant lot management in 1995. Having had experience in vacant land reclamation, the PHS in the 1990s was commissioned to produce a report on best practices for vacant land management. The report's recommendations included these: land acquisition and title consolidation procedures should be streamlined; policies and procedures should be established to facilitate community use of land; funding should be provided for brownfield remediation; and a community-based open space management program should be created.[1]

As a demonstration of this open space management project, the PHS launched the New Kensington Project in 1996, in partnership with the New Kensington CDC (Russ 2012). Funds for the initiative came from the Office of Housing and Community Development, the William Penn Foundation, Pew Charitable Trusts, the Local Initiatives Support Corporation, and the Philadelphia Urban Resources Partnership. New Kensington was identified as a high-priority neighbourhood, having suffered from dramatic population loss (from 32,000 to 15,600 people in forty years) leading to 1,100 vacant lots. Within three years of this project's launch, community members had reclaimed 370 parcels, built 62 gardens, cultivated 158 side yards, and established a demonstration garden and educational programs. The PHS hailed this project as an example of the potential for not-for-profit and for-profit urban agriculture in neighbourhoods with vacant land (Lawson 2005).

In the early 2000s, under Mayor John Street, blight removal became a key strategy for urban revitalization in Philadelphia. This effort, formally named the Neighborhood Transformation Initiative (NTI), had a budget of $295 million for building removal and land acquisition. The removal efforts resulted in around 31,000 vacant lots, which the city struggled to maintain. Estimates were that it would cost $57 million to convert all of the vacant lots into seeded and grassy spaces (Francke 2003). That level of funding was not available, so the city allocated $4 million to green 1,000 key properties around Philadelphia and hired the PHS to carry out the work. Half the funds were earmarked for debris removal, fertilizing, seeding, and fencing in 1,000 lots in North Philadelphia, Frankford, Mount Airy, north-central Philadelphia, West Philadelphia–Mantua, and South Philadelphia. One million dollars was committed to converting some properties into parks and gardens (Fleming 2003).

The success of vacant land reclamation and conversions, paired with the real estate boom in the early 2000s, led to the rise of public–private partnerships aimed at urban revitalization. The city's goal was to attract developers and private investment to the vacant lots. Councilwoman Jannie Blackwell was quoted in the *Philadelphia Inquirer* as a strong supporter of the program: "I think the private developers would do all of [Mantua] if we give them the properties. They are only held up if we are held up. They're ready. They are willing" (Francke 2003). Supporters of the NTI believed that the key was to make redevelopment easy for developers; eliminating blight and making land more attractive was part of this strategy. In addition, the NTI enabled the city to consolidate vacant lots into larger lots, something that might pique the interest of developers seeking large-scale projects.

To attract private investment and developers to neighbourhoods with extremely high vacancy rates and high numbers of vacant lots, the city sought to beautify these neighbourhoods. Mantua, in West Philadelphia, was an early target of the NTI demolition program. It had around 827 vacant lots (from a total of 3,027 parcels) (Francke 2003). In the early 2000s, a real estate boom led to commercial and residential investment in many neighbourhoods. Developers expressed interest in Mantua: the non-profit Neighborhood Restoration announced its intent to build two hundred homes, and Pennrose Properties proposed to build between fifty and sixty rental housing units (Francke 2003).

The city viewed vacant lot stabilization and greening as a means to catalyse large-scale capital improvement projects. But not all residents agreed with this strategy. Many community members viewed the gardens on vacant lots as integral to neighbourhood and community life, not as a temporary use of underutilized urban space. But in the 2000s, just as in the late 1920s, vacant land was becoming more and more attractive to private investors and developers. In the 2000s some community gardens were "paved over" by residential and commercial buildings, much to the frustration of many gardeners, who had poured volunteer hours into maintaining and sustaining them. For instance, the 2100 Fitzwater garden in the Southwest Center City neighbourhood was lost to development in the 2000s . Gardeners documented the loss of that garden on a website, *South Philly Blocks*. It had been established in 1994 and maintained by local volunteers, who applied for and won grants to develop and maintain it. Local volunteers had tended it, including children and the elderly. The garden provided a location for community activities and gatherings.[2]

As the real estate market started to pick up in the early 2000s, developers began looking for more parcels to develop. Southwest Center City, where 2100 Fitzwater was located, was becoming an attractive location for developers and new residents because of its proximity to the business district and its lower-cost homes. Gardeners posted a description of the garden's demolition in 2002 when a developer purchased the parcel the 2100 Fitzwater garden occupied. The garden was levelled a week after the sale. Strong controversy surrounded the sale of the property and the demolition of the garden. Neighbours reported that they had received little notification of the sale and that once the sale was complete, they had little time to negotiate or respond.

The gardeners' frustration highlights the sensitive land tenure politics surrounding the city-endorsed vacant lot cultivation strategy. The city viewed gardens as a temporary land use until development happened; the gardeners viewed them as permanent – indeed, as vital to their communities. On their website the gardeners emphasized the developer's ruthlessness and the city's lack of both compassion and help. Their accounts underscore the failure of agreements, both written and informal, to preserve any of their garden, including the plants they had planted and tended for several years. For instance, the subtitle of the webpage reads: "A signed agreement with the owner did not protect these gardens once the property had been sold." Furthermore, once the property had been sold, the gardeners discussed the sale with the buyer's lawyer and requested time to relocate the plants. According to the gardeners, that lawyer had assured them that demolition would not begin until the gardeners met, but this assurance was not respected, and the garden was soon demolished.

Additional excerpts from the website highlight the scope of the problem in the context of real estate law. Neighbours felt betrayed by the landlord's decision to sell the property and were further frustrated by the false assurance that their garden would not be demolished – an assurance that was broken through a legal loophole (Blanchard 2005). That statement highlights the vulnerability of urban gardens. The 2100 Fitzwater garden was well supported by many local residents, but that local support could not defy real estate laws or defeat the lawyer who had been hired to facilitate the parcel's redevelopment. The exchange value of the land trumped the use value of the garden, and the gardeners did not have sufficient legal protection to assert their claim.

Finally, the neighbours highlight the irony that garden development can lead to garden demolition: "Green spaces such as the 2100 Fitzwater Street Garden have the potential to transform neighborhoods without additional investments of public funds.

Ironically, it is probably the efforts of the gardeners of Fitzwater Street that played a key role in the real estate boom that heralds this garden's possible demise" (Blanchard 2005). The replacement of the garden with a new house and the lack of attention to local gardeners' wishes underscores how local laws give urban gardens a temporary status, even when those gardens serve as anchors for community activities.

Fitzwater was one of many displaced gardens in Philadelphia, and its disappearance exposed a perennial conflict between gardeners and other urban land use interests. Gardens help social institutions accommodate crises (Bassett 1979), but historically they have not been incorporated into official long-term planning for economic growth. In some ways, this tension is similar to what was experienced in the 1920s, when land for gardening became scarce and the PVLCA found it difficult to secure land for gardening each year. In the 2000s, however, gardens were more than sites of food production. The gardeners viewed their spaces as community resources and neighbourhood assets. From the perspective of developers and the municipal government, gardens helped eliminate blight and made the land they were on and the neighbourhood they were in more attractive for tax-generating development. This disagreement underscores the vulnerability of urban gardens over time: the reclamation of vacant lots is a pressing need for Philadelphia, but the various stakeholders (residents, city councillors, the city administration, etc.) do not always agree on *how* to reclaim them.

In the past, these land disputes usually ended with development demolishing or displacing the garden. However, expectations have been changing in this regard. In 1986, Philadelphia's first garden trust was established, called the Neighborhood Gardens Association (NGA); more than thirty gardens have been entered into the land trust since its establishment.[3] That this program has survived speaks to the public's changing perception of the importance of gardens. The land trust was not part of official policy; even so, it was able to step in to address a community need for urban gardens by ensuring their presence in the city in perpetuity.

Garden land trusts mitigate the vulnerability of gardens during times of development and increasing demand for land.

Gardening as Social Service for Youth

Besides managing open space, gardens have served other societal initiatives. Many accounts of urban gardening describe the experiences of youth involved in gardening projects as a wholesome and healthy outlet (Pudup 2008). As with the vacant lot maintenance agenda described above, youth gardening programs have evolved through the years, expanding their potential impact in terms of scope, scale, and audience. Early programs featured gardening as a self-help opportunity for so-called troubled youth; more recent iterations of these programs emphasize the job training that youth gardening programs provide their "employees," as well as their potential to address chronic unemployment in neighbourhoods with high vacancy rates. These multiple agendas embrace the moral and economic imperatives of the early gardening movements, as well as the community-oriented "grassroots" empowerment efforts of the 1970s.

In the 1990s, some youth workers started to promote gardening as an option for youth growing up in low-income neighbourhoods exposed to crime and drugs. Some proponents of youth gardening programs felt that working outside in the soil would allow students to reconnect with nature, besides fostering a sense of land ethic: "By nurturing the seeds and tending the garden, the students learn not only basic botany, but also stewardship of the Earth and self-esteem" (Campbell 1998).

Youth gardening programs often serve children from low-income and underprivileged backgrounds. Especially in a context of economic depression and loss of social services, advocates for these programs claim that gardening provides prompt and tangible benefits, including the pleasure of seeing the first sprouts of a seedling, the satisfaction of digging in the dirt, the rewards of selling produce, and the benefits of learning entrepreneurial skills. The following quotes from the *Philadelphia Inquirer* from the late 1990s and early 2000s demonstrate how these youth gardening programs and their benefits have been framed as a remedial activity for urban youth:

> "For a lot of them, it's hope, especially in a city where we don't always see a lot of positive outcome for what we do." (Gordon, 1996)

> "We're trying to teach the youths how to do something positive ... When you see the stuff you planted starting to come up, that's a different experience. They'll feel good." (Campbell 1998)

> "The whole idea was that the kids could be empowered to grow their own food ... they are empowered so they can always rely on themselves." (Cowie 2001)

Clearly, youth gardening programs are meant to be positive, empowering, and rewarding. These programs see gardening as a way to motivate youth, possibly capturing their attention in ways that after-school programs might not. This implies that gardening is unusual in the urban context and that youth will find it novel and exciting to cultivate plants. Also, the experience of gardening in the city can have spillover effects and teach youth important and transferable life skills.

While these youth gardening programs share some of the moral objectives of the early vacant lots programs from the 1900s, many of them also have a mental health orientation, as observed in Pudup's (2008) study of organized garden projects in California. Youth gardening advocates promote gardening as a form of therapy, especially for young people with challenging childhood experiences and few places to turn for assistance:

> Through the [Veggie Kids] program, "we build and direct culture," says Hussain Abdulhaqq, 30, a community gardener, who created the North Philadelphia project, and germinates all the beds. With the program for seven years, Abdulhaqq says he's seen grades go up and behavior improve. He's noticed how troubled ones use the garden as an outlet.

"Here they can learn to deal with their problems better than they could on the street," he says. (Gregory 2009)

The discourses on youth gardening highlight the benefits of gardening compared to other popular activities for children and teens, such as video games and watching television. Advocates of youth gardening programs argue that gardening provides a unique social service for youth, by creating opportunities for healthy eating and exercise and for learning responsibility and other life skills, including a land ethic. Proponents of these programs typically describe them as a means to an end: youth development. They also argue that there is something unique about working the soil that is not captured by other types of youth activities.

Recent concerns about "food deserts" and the lack of access to fresh and healthy foods in low-income neighbourhoods have added a nutritional objective to many youth gardening programs. The US 2008 Farm Bill defines a food desert as an "area in the US with limited access to affordable and nutritious food, particularly such an area composed of predominately lower income neighborhoods and communities."[4] Many communities have bodegas or fast food restaurants that sell packaged foods, but few stores or stands that sell fresh produce, which is healthier than processed food. Residents seeking fresh produce might be unable to travel to a grocery store, or unable to afford such items. In many North American cities, food deserts are the primary concern when it comes to food security.

Urban gardens have been implemented by Philadelphia's *Greenworks* plan as one way to address food insecurity in low-income neighbourhoods (Nutter 2009; and see chapter 4). Youth gardening programs are being promoted as a way for youth to involve themselves in a community-based urban garden, while simultaneously tasting fresh produce and learning healthy eating habits.

More recently, youth gardening programs have promoted the economic benefits of gardening by involving youth in the sale of produce; this promotes financial literacy and business skills. Organizations like Teens 4 Good (founded in 2005), the Walnut Hill Community Farm (opened in 2010), Mill Creek Farm (started in

2005), and the Urban Tree Connection (founded in 1989) employ youth to plan the planting season, maintain the farm, harvest the produce, and sell this produce at farmers' markets around the city. One program manager in Philadelphia explained how business skills were incorporated into their youth program: "The big thing for this year that we are emphasizing more than last year is the financial literacy aspect of it. They are learning how to keep ledger books. How to assess your stock and do inventory" (Interview 4). This organization's focus on financial literacy emphasizes how gardening can teach youth transferable skills. It also acknowledges that youth gardening programs can help prepare youth for jobs other than farming. Another program manager highlighted how the job skills learned through urban gardening were useful beyond the program and that the program could help youth see potential career paths that might have seemed unattractive or out of reach:

> It's showing them skills in horticulture and landscaping, it's giving them entrepreneurial skills, it's giving them a lead to a career path they may never have thought of … The crew that is working with the [grocery] stores, they're getting a chance to learn that there's a human resource, what does the produce manager do? There's a whole behind the scenes, so they're starting to learn that. (Interview 10)

This increasing focus on economic opportunities and viable employment options reflects a new trend in gardening. As noted earlier, youth gardening programs were initially promoted as therapy for at-risk youth. Gardening programs had the potential to keep kids out of trouble. Today, however, more organizations are pitching these programs as presenting new opportunities for motivated youth from underprivileged backgrounds. Youth can join a summer gardening program and develop entrepreneurial skills and possibly identify job prospects in landscaping or other related fields.

The emphasis on economic skills represents a shift from the focus on mental health. And it reflects another trend. Gardeners are seeking ways to make urban agriculture a permanent part of the urban landscape. If urban farms can gain a foothold in a sustainable urban food system, they will be less vulnerable to displacement. By adding a youth entrepreneurship option, gardens provide an extra benefit; also, such programs can bring gardens additional funding and resources to make them viable over time. At a time when many gardens struggle to remain sustainable from year to year from produce sales alone, youth entrepreneurship programs enable urban gardeners to diversify their gardens' activities and thereby justify their continued existence.

This vision of gardening as a long-term, paid activity reflects a new optimism among urban farmers and their advocates. It also reflects an effort to professionalize urban agriculture, beyond youth programming. In 2012, more and more farmers were pursuing urban farming as a career choice. Some farmers were recent college and university graduates who had opted to involve themselves in urban gardening in a formal capacity, beginning as apprentices and later starting their own urban farms. Some youth programs emphasized this career path to their youth; others focused on the transferrable skills that youth could gain by participating in summer gardening programs. In a later chapter we revisit the career pursuits of urban farmers and consider whether there is enough support for urban gardening to sustain a new class of career urban farmers. A parallel theme in gardening is the growing green identity in Philadelphia: gardens are becoming integral to the city, even if there is little consensus on their specific role.

Gardening as Part of Philadelphian's Green Identity

Thanks to the efforts of the NGA, the PHS, and the PSU Garden Program, urban greening efforts have become increasingly part of Philadelphia's urban imaginary. Of course, this imaginary is as diverse as that city's residents. It runs the full spectrum, from the garden as an isolated oasis in a gritty city to the garden as the future identity of a green Philadelphia.

The following quotes from the *Philadelphia Inquirer* present the ways in which Philadelphians have embraced the "green city" concept as part of their city's image. In the early days of economic decline, green gardens were a place to escape the gritty city for a short time. Over time, however, more and more residents and planners have come to see a green Philadelphia as the city's way forward, as part of a new, sustainable Philadelphia. One of the densest cities in North America, Philadelphia long lacked green space in many neighbourhoods. By using gardens as signs of environmental sustainability rather than as a reflection of abandonment and limited fiscal resources for maintenance, the city of Philadelphia, with the support of many of its residents, has created a new vision for a greener city.

Philadelphia counted its highest numbers of vacant lots in the 1990s, and descriptions of community gardens that appeared during the most dramatic periods of urban decline described them as means of escape. Accounts of gardens in abandoned neighbourhoods and vacant lots conjured up images of reprieve from urban decline:

> [Garden of Eatin'] is an oasis, surrounded by crumbling houses, trash-strewn lots and an abandoned mayonnaise factory. Each garden plot is divided by white picket fences. It is here that neighbourhood gardeners, most of them elderly, nearly all from the South, come to escape urban blight and reconnect with their youth. (Vitez 1994)

The above account, from the mid-1990s, suggests that Philadelphia has been lost to abandonment, crime, and dilapidation and that gardens allow gardeners to imagine themselves to be somewhere else for a short period of time.

Others described garden development as a way to reclaim devastated neighbourhoods. Crime rates in the most abandoned and neglected neighbourhoods in Philadelphia rose in the 1980s and 1990s, and the drug trade made many residents afraid to venture outside. Planting gardens was one way for communities to reclaim vacant land and establish safe spaces. The following report from the *Inquirer* about an Urban Tree Connection garden in West Philadelphia's Haddington neighbourhood suggests the transformative power the garden had for some residents:

"It was a barren place," says Lisa Barkley, who has lived in Haddington Homes almost all of her 52 years.

Before it was a garden, Pearl Street, long closed to traffic, residents say, was a shadowy alley of weeds, trash, abandoned cars, broken glass, prostitution and drugs.

A band of neighbors proved to be more stubborn.

"We took it back," says Barkley, standing over the vegetable and herb garden she planted last year.

"Our goal is to make this a nice haven, where kids can play safely. They" – the drug dealers – "know how serious we are, so they won't come back." (Gregory 2009)

In this account, it seems clear that a garden can have a positive impact beyond its own borders. Here, a garden serves as both a community anchor and a change agent. Just as important is the community's role in instigating neighbourhood transformation. This account suggests that residents are capable of revitalizing their own community and that gardens can play an important role in that process. The PHS has attempted to facilitate this sort of community-led neighbourhood revitalization since the 1970s through its Philadelphia Green efforts and also through the more recent Garden Tenders program. The latter provides gardening training and experience as well as information on finding a suitable site in Philadelphia; it also identifies resources to maintain the garden and offers tips for assembling a supportive community of growers. Through these programs, the PHS aims to create a citywide community of growers who pursue and maintain their gardens independently. The PHS does not maintain the gardens through their organization; it does provide technical assistance and resources for gardeners.

The community gardening movement arose in times of strife, with government agencies and philanthropic organizations providing low-income families and unemployed labourers with resources and technical support. But interest in gardening was not limited to low-income residents. Many other Philadelphians found urban gardens attractive and gardening a compelling activity. The 1990s saw a new wave of interest in urban gardening that transcended more demographic boundaries than in earlier gardening eras. As evidence of the impact of greening on the urban imaginary, an article published in the *Inquirer* in 1994 boasted that gardening was infectious (Heavens 1994):

> The gardening bug does not discriminate. It even bites folks who have front yards made of concrete and whose back yards are narrow alleys, fire escapes or flat, asphalt roofs.
>
> Stroll along any residential street in Center City and you'll soon discover that the bug has found plenty of willing victims – all people who don't believe that a neighborhood's green spaces should be restricted to weeds growing between the cracks of a sidewalk.
>
> Dotting the neighborhoods are trees bordered by small wire fences or wooden boards that protect impatiens of all colors from curious animals. Half-barrels full of bright red geraniums line the sidewalks. Vinca and ivy hang from window boxes of every design, shape and size along streets in Fairmount and Spring Garden, in Society Hill and in Northern Liberties, and on the tiny streets and alleys amid downtown skyscrapers.

By the 1990s, gardening was becoming part of everyday life both in bustling neighbourhoods and in areas battling vacancy and abandonment. Many of these gardening activities were notably evocative of a past time – William Penn's notion of the Greene Countrie Towne. To be more precise, gardening proponents described the movement as a "second chance" to fulfil William Penn's dream of a green Philadelphia. Thus, the gardening movement became a way for Philadelphia to re-create its image as a green city and to justify a lower-density city.

Advocates for a green city recognized the challenges of gardening in an older industrial city like Philadelphia. The close-packed buildings and narrow streets did not reflect what many people thought of as an optimal urban layout, particularly for gardening. One account of Philadelphia describes the creativity required of gardeners because of the constraints posed by the built environment:

City gardeners tend to be a hardy lot, mainly because cultivating urban areas requires hard work and a huge amount of creativity. Because many city houses were built close together, homeowners often are forced to hoe areas no bigger than postage stamps. Not only is the space limited, but the shadows of adjacent rowhouses and high-rise buildings often shut out the sun. (Heavens 1994)

Despite the challenges posed by the built environment, more and more people, including residents, city officials, and members of gardening associations, turned to gardening as a way to think about the future of Philadelphia. During the abandonment and high crime rates of the 1990s, many people looked towards the rebuilding of the city. One journalist described the gardening efforts as a way of "taming and beautifying" the city (Sokolove 1994). The vision of a green city instilled hope that a future Philadelphia would attract people and provide a high quality of life. Philadelphia's era as an industrial hub was waning; greening initiatives indicated a transition towards a different type of city.

Gardening as the Vision of a Sustainable Philadelphia

While the gardening movement held promise for the rebuilding and revitalizing of Philadelphia, the vision for a future Philadelphia was not clear or agreed upon by the various stakeholders that were planning Philadelphia's future. One aspect of the change was clear – Philadelphia was becoming less densely populated. As Seplow (1999) pointed out, Philadelphia's population loss from 1950 to 1999 approximated the size of the nation's capital. Blaine Bonham from the PHS commented ironically in the *Inquirer*: "This has come full circle from the little open space to the huge open space" (Cowie 1999). He saw this open space as an asset for the city. According to proponents of a green city, Philadelphia had little park space compared to other cities of similar size (Harnik 2010).

Clearly, something needed to be done to address the extensive vacancy in Philadelphia. Views on this matter, though, ranged from accepting to horrified. Loeb (1994) described several different opinions of the demolition in Philadelphia (i.e., those of a preacher, an architect, and a housing director). A preacher from North Philadelphia associated the tearing down of old buildings and the building of new neighbourhoods with the Book of Genesis and new beginnings. An architect and preservationist from Center City called the demolition "architectural genocide," preferring to renovate and rebuild Philadelphia rather than attempt to treat certain neighbourhoods as a blank slate. In a very different view, the city's housing director proposed the semi-suburbanization of the city – that backyards and driveways be added to newly built homes to replace the dilapidated and ultimately demolished homes.

Most people in the 1990s did agree that Philadelphia should plan for reduced density. The director of the City Office of Housing and Community Development acknowledged the reality of the population loss in the local paper: "We can't replace all the houses because we need fewer homes" (Seplow 1999). Depending on who you spoke to, the transition of Philadelphia could lead to anything from large side yards for more individuals to parking pads or other types of development. However, the opportunity to create more gardens or parks was a constant in these discussions. By 2009, with the release of the city's formal sustainability plan, city officials were starting to view the vacant lots and their gardens as potential assets in Philadelphia's transition to a sustainable future

Gardening has long been present in the city, and today gardening seems to be as popular as ever. The combination of this renewed interest in gardening and the city's support for institutionalizing gardening suggests potential for some new form of urban gardening – formal, permanent, or something else. While gardening continues to be a debated and (in some cases) contentious topic for various stakeholders, many gardeners today are excited about this moment. Something seems new or different when it comes to cultivating urban plots. The next chapter explores the burgeoning efforts to connect urban agriculture with urban sustainability planning.

chapter four

The Politics of Urban Agriculture in Sustainability Planning

If urban agriculture is, as many have argued, a critical component of a sustainable city, it needs city support and to be recognized in city policies. This chapter examines the processes of promoting urban agriculture through demonstration projects that receive city support as well as through systematic policy reforms. As noted in the previous chapters, until recently, even though there was interest in urban agriculture, the city primarily left it up to the citizens and to third-sector NGOs. For people interested in urban agriculture, there were, of course, challenges that came with the city's hands-off approach. That approach meant that it was difficult to access water and that the growers did not have land tenure, particularly if they were squatting on vacant lots. As previously described with regard to 2100 Fitzwater, gardeners could be kicked off the land whenever a developer decided the time was ripe to redevelop it. There was no clear path to land ownership and protection. Changing this dynamic has required commitment on the part of city officials as well as active citizen engagement and is an evolving process. Important shifts include the Philadelphia Water Department's interest in promoting urban farms in the early 2000s and then the Nutter administration's recognition of the role that urban agriculture could play in supporting a sustainable city agenda by helping to provide fresh food in underserved neighbourhoods. Under Mayor Nutter, urban agriculture became a part of the city's policy agenda with the adoption of the City Food Charter, the inclusion of urban agriculture in the *Greenworks* plan, rezoning to allow for urban agriculture, the development of a Food Policy Advisory Council (FPAC) (staffed by the city's Office of Sustainability), and the development of more streamlined city land disposition processes combined with the creation of the Land Bank and closer city partnerships with non-profits such as the PHS and the Neighborhood Gardens Trust (NGT). Importantly, the city's efforts have been supported by NGOs and citizen-activists, who have mobilized to demand a more inclusive approach to policy-making to support urban agriculture. Their activism has kept urban agriculture on the urban policy agenda, helping solidify the idea that a sustainable city includes opportunities for growing in the city as well as mechanisms for community input and leadership in planning processes. As we highlight below, urban politics, existing governance and decision-making arrangements, and competing demands on city resources and land often complicate efforts to formally promote urban agriculture. Even so, important progress is being made in Philadelphia that may offer lessons for other cities that are wrestling with the question of how to integrate urban agriculture into citywide decision-making.

City Efforts to Promote Urban Agriculture on City-Owned Land in the 2000s

To connect urban agriculture with urban policy-making, there needs to be support for the notion that urban agriculture is an accepted and valued use. In Philadelphia in the early 2000s that leadership came from officials within the Philadelphia Water Department (PWD), a city agency that has been at the forefront of rethinking the relationship between engineered and natural systems. The PWD has been nationally recognized for its *Green City, Clean Waters* plan (adopted in 2011), which promotes the use of green infrastructure (bioswales, rain gardens, green roofs, permeable pavement, urban farms, etc.) to capture the first inch of rainwater throughout the City of Philadelphia to keep it out of the city's combined sewer system (which combines sewage and stormwater). By using green infrastructure to manage stormwater, the city hopes to prevent the overflow of the city's combined sewage system during rain events into the Schuylkill and Delaware Rivers; this would reduce the need to invest in expensive pipes and treatment facilities (grey infrastructure) in order to meet the Federal Clean Water Act requirements.[1]

Even before the adoption of *Green City, Clean Waters*, the PWD's leadership had a strongly progressive view of urban agriculture. They saw community, environmental, and economic benefits to promoting urban agriculture on city-owned land and to coupling urban agriculture with the city's goal of promoting green infrastructure. PWD officials argued that we should support a green approach to stormwater management because green infrastructure has social, economic, and environmental benefits; this is particularly important for a city like Philadelphia, which faces high vacancy rates and a lack of access to green space and low-income access to fresh food.[2] Urban agriculture fits in well with this more holistic vision of the city. In the early 2000s, PWD officials saw opportunities to connect some of the exciting grassroots work being done by urban gardeners and farmers with the city's goals of managing urban stormwater through urban greening. PWD officials also saw the potential economic benefits of commercial urban agriculture and wanted to test the feasibility of promoting urban farming on city-owned land.

In 2003, Nancy Weissman, the PWD's then director of economic development, was quoted in the *Philadelphia Inquirer* advocating for commercial farming. She asked, "Is a farm economy possible in Philadelphia?," and then answered, "We believe it is."[3] In 2003 the PWD leased a half-acre of land to the Institute for Innovations in Local Farming (IILF), a non-profit, to develop a commercial urban farm at its Somerton Tanks Farm property in Northeast Philadelphia using the SPIN (Small Plot Intensive) farming method. The project was an experiment, funded by the PWD, city, state, and federal agencies, and non-profits, to test the viability of commercial farming in Philadelphia.[4] In 2005, through a collaboration with the PWD and the Pennsylvania Horticultural Society (PHS), with funding from the Pennsylvania Department of Environmental Protection's Growing Greener Grant, the Redevelopment Authority

(RDA) granted a ninety-nine-year lease to Mill Creek Farm in West Philadelphia for vacant land where a row of houses had collapsed due to settlement issues from a buried creek under the property.[5] Since then, Mill Creek Farm has been a highly successful project demonstrating the potential synergy between the PWD's goals and other community, economic, and environmental benefits.

Development and Preservation of Mill Creek Farm: The Impact of Councilmanic Prerogative

The Mill Creek Farm has been a leader in the city in providing youth environmental education, food access for low-income people, and opportunities for community engagement; but even though it is supported by city agencies, its ninety-nine year lease can be broken with thirty days' notice. As a result, Mill Creek Farm, like other farms and gardens in the city, has existed in a state of perpetual uncertainty about its long-term land tenure. Urban agriculture activists have continued to lobby the city councillor for West Philadelphia (where Mill Creek Farm is located), Jannie Blackwell, to place Mill Creek Farm and the neighbouring Brown Street Community Garden in a land trust through the Neighborhood Gardens Association (which became the Neighborhood Gardens Trust in 2012) to guarantee that both will be able to stay where they are. According to a policy brief by Jade Walker, co-founder and former co-director of Mill Creek Farm, the farm and the garden gathered more than nine hundred signatures and had more than ninety letters of support for protecting Mill Creek Farm from future development.[6] However, Blackwell saw the support for Mill Creek Farm as coming from outsiders, whom she described as "people who do farms and move around the City or wherever they move to, but I don't know them."[7] Instead, Blackwell saw the farm as a potential site for a housing development (despite settlement issues that, urban agriculture advocates argued, made the ground unsuitable for redevelopment).[8] During a City Council meeting on 16 June 2010, Blackwell said she hoped to see a housing development on the Mill Creek Farm site and would be open to moving Mill Creek Farm three blocks away to another vacant parcel in a land swap.

(The new site, while larger, would face even more settlement issues, so she argued that it would make more sense as a farm.)[9] In her statement to City Council, she expressed some concerns about Mill Creek Farm, describing it as "weedy" all winter long, and argued that she had a right, under her councilmanic prerogative – whereby city councilors effectively have the final say on many land use decisions, particularly on city-owned land in their districts, because other city councillors will defer to their decisions – to advocate that the site be used for much needed housing.[10] She argued that her proposal to relocate the farm and build houses on its original site would allow the development of housing *and* a farm. Blackwell asserted her "right to fight for housing in my area. I have a right, when I can have housing and a garden within three blocks I have a right to be able to say that."[11] Advocates of Mill Creek Farm were disappointed by Blackwell's position and frustrated that they were unable to protect Mill Creek Farm through a land trust. They saw the reluctance to protect urban agriculture as an indication that some city councillors still do not accept urban agriculture as a "highest and best" use of land; with regard to Mill Creek Farm, they thought this was ironic, given that the site has settlement issues and is probably unsuitable for development.[12] But Blackwell's opposition to the Mill Creek Farm does not mean she is opposed to protecting all urban gardens and farms in the city. While she has not yet agreed to preserve Mill Creek Farm, she was a champion of the Saint Bernard Community Garden, protecting it from sheriff sales in 2012 and 2016 and supporting its transfer to the Neighborhood Gardens Trust.[13] In 2016, when asked why she supported protecting the Saint Bernard Community Garden, she explained to *Plan Philly* how important the garden was to the community: "I better intervene. You know how long these people have been working for these gardens?"[14] Many growers we interviewed across the city were frustrated that their beloved community gardens and farms were ultimately at the mercy of what they saw as the "whims" of their city councillors. They wanted a more systematic approach. However, some city officials and citizen-activists defended the use of councilmanic prerogative and decisions like Blackwell's (to support gardens and farms in some cases and not others), arguing that city councillors understand the broader is-sues impacting their districts: they have a more nuanced under-standing of community needs than the urban growers, some of whom are not from the communities where they want land for gar-dens and farms (this is particularly true of young, white farmers). City councillors are democratically elected and may also be aware of developer or city agency interest in particular parcels of land.[15] Also, some local residents may perceive urban agriculture as a less beneficial land use than residential or commercial development. In cases where a district is primarily African American and the urban farmers are not from the neighbourhood, city officials and activists who reject requests for urban agriculture may in fact be protecting that neighbourhood from what may be perceived as an "invasion" by young millennials, who may be contributing to increased gen-trification pressure (discussed more in later chapters).

Greenworks and the Institutionalizing of Urban Agriculture

By the time Michael Nutter was elected mayor of Philadelphia in 2007, there was already a buzz that urban agriculture might be a solution to many of Philadelphia's problems, such as vacancy, lack of employment, lack of green space, stormwater management, and food deserts. In part, this buzz was a reflection of the PWD's lead-ership, the city's long history with neighbourhood gardens, and the above-mentioned projects like Somerton Tanks Farm and the Mill Creek Farm. But it was the creating of the Mayor's Office of Sustainability in 2009 that launched a new era of citywide official support for greening initiatives and urban agriculture. Initially, ur-ban agriculture connected to the city's sustainability agenda pri-marily through the desire to improve local communities' access to fresh and affordable food. In 2008 the city adopted the Philadel-phia Food Charter, which pointed specifically to "urban agricul-ture policy" as a means to achieve the city's food access goals. The Philadelphia Food Charter presents a vision for a food system which benefits our community, our economy and our environment and helps push Philadelphia further towards becoming the Greenest City in America. It establishes the City of Philadelphia's commitment to the development of a coordinated municipal food and urban agriculture policy, and articulates the intention of establishing a Food Policy Council populated by key city and regional stakeholders who can inform and advise the city's efforts while helping to provide coordination, momentum and support for the significant activities already underway throughout the city and region."[16]

The adoption of the Food Charter and the subsequent incorporation of many of its food goals into the *Greenworks* plan, the City's 2009 sustainability plan, was a win for activists interested in urban agriculture and food justice. *Greenworks* specifies many sustainability targets addressing environmental, economic, equity, and management-related issues, including equitable access to local fresh food and recreation opportunities. One of *Greenworks'* equity goals is to "bring local food within 10 minutes of 75 percent of residents."[17] *Greenworks* also calls for the creation of "twelve commercial agriculture projects over the next 8 years" and emphasizes the leveraging of city-owned vacant land in order to promote economic opportunities and increase low-income access to fresh food.[18] These goals address concerns that disadvantaged neighbourhoods in Philadelphia are "food deserts" in that they lack access to a supermarket or other outlets for healthy food.[19] Urban agriculture is seen as a solution, one that promotes access to affordable fresh produce in the city, as well as green jobs.

However, while *Greenworks* formally acknowledged the importance of urban agriculture and its role in greening the city, the promotion of urban agriculture in Philadelphia has been hampered by fragmented municipal governance and by the city's outdated land disposition processes, which have made it extremely difficult to assemble and control land. This has been particularly frustrating for urban growers, who identify vacant parcels suitable for urban agriculture only to find that the identified site is actually owned by several city agencies (and sometimes also by multiple private property owners), each with its own land distribution process. And even if they are somehow able to navigate the complicated land acquisition processes with those multiple owners, access to vacant land can be blocked by city councillors under councilmanic prerogative because the city councillor for the district where the land is located makes the final determination on the transfer of city-owned land (as discussed earlier in the Mill Creek Farm example). An often-heard complaint in 2011 from urban growers was that while *Greenworks* called for urban agriculture, the plan did not have the "teeth" it needed to implement a comprehensive urban agriculture policy. An urban grower expressed frustration with how difficult land access was:

> Nothing can happen without Council approval in a neighborhood like ours where we have a Council member [name withheld] who does not give a damn about community gardening. So nothing happens unless you do it yourself or you do it illegally or whatever. So we can talk about *Greenworks* and all this other sh#?, but there's no teeth. (Interview 3)

Greenworks acknowledged the value of urban agriculture, yet residents and activists who were excited about getting their hands on a vacant lot continued to find themselves navigating a maze of city agencies just to determine who owned the land. Gaining legal access to it was exceedingly difficult: some felt that the problem was a lack of will at City Hall; others felt it was a lack of resources. They were frustrated that *Greenworks* did not give the Mayor's Office of Sustainability the authority to counter many of the long-standing barriers to urban agriculture such as the councilmanic prerogative, the prohibitive cost and difficulty of accessing water, and an antiquated city land disposition process.[20]

Including Other Voices in Decision-Making: FPAC, Soil Generation, and the PHS

To help implement the Philadelphia Food Charter and promote the goals of *Greenworks*, the city created the Food Policy Advisory

Council (FPAC), consisting of local officials and activists and staffed by the Office of Sustainability. It was modelled after food policy councils in other cities. FPAC has been an important part of engaging with a more diverse group of citywide stakeholders to influence policies around food and urban agriculture.[21] The founding of FPAC was a significant policy development for urban agriculture that many cities will want to replicate; however, in Philadelphia the development of a council that represents the city's diverse interests and demographics has been challenging. Created in 2011, FPAC started out with twenty activists and city leaders appointed by the mayor.[22] The group was slow to ramp up, in part because, since the members would be playing an advisory role to the mayor, they needed to be carefully vetted (to ensure that they were qualified to play an official role). Once established, FPAC started by examining citywide sustainability and planning efforts (in addition to *Greenworks*) – particularly *Green 2015*, Philadelphia 2035, and Eating Here (the Delaware Valley Regional Planning Commission's regional food plan) – to identify ways it could help promote access to healthy and local food. The council created subcommittees to focus on different issues: urban agriculture (originally it focused on vacant land), anti-hunger, food and health, zero waste, and workforce and economic development.[23]

Since 2011, FPAC and its subcommittees have worked to develop policies – for example, the inclusion of urban agriculture in the city's rezoning, the development of the Land Bank, and soil safety policies to meet the goals of *Greenworks* (discussed below). However, the process of bringing adequate representation to FPAC has not always been smooth. While that body consists of representatives from throughout Philadelphia (urban farmers, community gardeners, activists, members of nonprofits, and city agencies), initially some urban growers did not feel represented by FPAC, nor were they on board with its policy proposals. They described it as "being top-down, not representative of people, either eating or growing food in the city" (Interview 15). One interviewee described the FPAC as "another example of white people deciding on a solution, creating that mechanism and then inviting people from the community after it is already started. People of color don't have agency within that space" (Interview 32).

FPAC has made significant efforts to be more relevant to the policy conversation and more representative of the City's diverse constituents. Between 2011 and 2017 it became a more established organization with a webpage, official membership, and dedicated staff. During this time there was also an increase in the number of people of colour on FPAC and its subcommittees.[24] But despite participation by more urban activists of colour, some interviewees felt that there needed to be more intentional mechanisms for people who are from disinvested communities to determine how community land is used. Figuring out an effective mechanism for engaging citizens in FPAC has been a significant challenge. Even the timing of its meetings can be problematic for growers and activists, many of whom wear multiple hats and juggle work and family responsibilities. Some noted that FPAC meetings are held from 3 to 5 p.m. on weekdays in a formal room in a city government building, making FPAC inaccessible and intimidating to many community residents.

A few interviewees also questioned how effective FPAC could be, given that it is advisory only and does not necessarily have political buy-in, particularly from city councillors (Interview 30, 32). The mayor and the City Council do not have to act on its policy recommendations, which may be perceived as not being core issues for their constituents, many of whom are African American. To better make the case that urban agriculture supports the goals of low-income and African American communities, one African American interviewee stressed the need for people of colour to assume leadership roles in urban agriculture and urban sustainability more generally:

> You do need a more effective bridge between communities of color and these efforts and that means that the folks who are in charge have to be willing to not be in charge. For the sustainability community at large, until it is inclusive, it won't be impactful. They need to be involved in the decision-making. (Interview 33)

Kirtrina Baxter is a black community organizer with the Garden Justice Legal Initiative (GJLI), a legal advocacy organization supporting urban gardens, and one of the founders of Soil Generation (formerly Healthy Food, Green Spaces), "a black-led, grassroots coalition of radical community gardeners and urban farmers working to build a hyper-local food system in Philadelphia that promotes health and equity in historically marginalized communities and works toward the creation and preservation of safe, healthy, economically secure and culturally-reflective neighborhoods."[25] She describes the important work Soil Generation is doing creating an alternative organization that focuses explicitly on social, racial, and food justice:[26]

> We are "browning urban ag." because we are challenging institutions to be accountable to the communities in which they work in. And also acknowledging that urban agriculture and growing in communities has been happening forever. Let's not forget the story of those folks who have been out here doing this work for such a long time. So we aren't just talking about food. We are also talking about self-determination. We are also talking about equity and what economic justice means in the context of our food system. How do people who have historically been displaced or not taken care of … how do we centre them in creating this new food movement? So that is new. (Kirtrina Baxter, 2 July 2016)

Soil Generation and the GJLI help growers with the legal and organizing tools they need to protect their gardens and farms and have a more powerful voice in city policy-making. Through a series of Train-the-Trainer workshops and the Vacant Land 215 Toolkit, Soil Generation and the GJLI have been instructing citizens on how to gain access to land for gardens and farms.[27] Along with other grassroots coalitions, the GJLI and Soil Generation have also pushed for more progressive citywide policy reforms, such as a focus on community land in the Land Bank and the recognition of urban agriculture as an as-of-right use in the zoning code (and protecting agriculture from efforts to limit it proposed by city councillors).

Increasingly, activists from these coalitions are taking leadership roles in organizations like FPAC: as the co-chair of FPAC, Amy Laura Cahn of the GJLI brought a social justice lens to FPAC and worked to better connect it with diverse constituents. Groups like FPAC, the GJLI, and Soil Generation are developing new avenues for policy dialogue about urban agriculture policy; established groups like the PHS are also collaborating with these groups to advocate for growers. The PHS now houses the Neighborhood Gardens Trust (NGT) (formerly the Neighborhood Gardens Association), and the PHS staff are leaders in FPAC bringing decades of experience working with communities and the city government on urban gardening.

City Support of Urban Agriculture Projects

Early city-led efforts to meet *Greenworks'* goal of supporting commercial urban agriculture demonstrated the complicated politics surrounding vacant land acquisition and community decision-making, even on city-owned land and even when the city took the lead. The requests for information (RFIs) for the Manatawna Farm, the Community Farm and Resource Center at Bartram's Garden, and the Marathon Farm highlight the various challenges facing city agencies when they try to promote urban agriculture.

Manatawna RFI: Lessons about Community Acceptance

To meet *Greenworks'* goal of developing larger-scale commercial urban agriculture, in 2010 the city's Department of Parks and Recreation tried to develop ten small, subsidized for-profit farms on a five-acre hayfield. The department's Manatawna RFI was for the development of ten half-acre chemical-free commercial farms that could sell to local farmers' markets. The proposal was to allocate five acres of land that was being used by the W.B. Saul High School of Agricultural Science, a public high school that focuses on agriculture.[28] However, the RFI was not well received by the surrounding community. The Manatawa RFI brought to light that not everyone wants urban agriculture in their community, that efforts to promote urban agriculture can get caught up in other city politics, including race and class politics, and that city councillors still have enormous control over their districts. Proponents of the project were surprised to encounter significant and highly vocal neighbourhood opposition to it, enough to ultimately block it. Neighbours of the Manatawna site (many of whom were white) argued that the proposed commercial farms threatened the "character" of the area and the local ecosystem. Several interviewees thought that the city had failed in its communication with the local residents, who were then able to frame the Manatawna farms as a threat to the existing use of the land when in fact the project was meant to develop synergies between existing neighbourhood assets such as the W.B. Saul High School and the small-scale, chemical-free commercial farms. Some neighbours claimed that the farm would threaten the local bird population; however, many urban farmers felt that other socio-economic issues were at play. The farms were to be located in a wealthier and primarily white section of the city, and larger half-acre lots would have been rented out at a subsidized rate to urban farmers so that they could grow enough to sell food.

Opponents of the Manatawna RFI questioned why the city was investing in urban agriculture in a place that was already in production and that neighbours did not view as "blighted." According to critics of commercial farm development at Manatawna, critical wildlife would be disrupted and the W.B. Saul School would have less farmland to grow hay for its livestock. One opponent of the project told the *Chestnut Hill Local* that "it's not just about raising vegetables – it's about benefiting the community beyond just a piece of food. With 40,000 vacant lots in the city of Philadelphia, why take a place that is already green and degrade it?"[29] This quote suggests that urban farming would actually "degrade" the Manatawna site. To stop the Manatawna Farms development, the local councillor, Curtis Jones, Jr, proposed an overlay district: the Roxborough Environmental Control. This would in effect prohibit commercial development *and* commercial farming on the land, except in areas already used for community gardening.[30] The overlay was unanimously approved by the City Council, effectively making the Manatawna commercial farms proposed by Parks and Recreation impossible. In the *Philadelphia Inquirer*, Jones defended his decision to prevent the development of Manatawna Farms: "Make no mistake, I am for urban farming, but not at Manatawna."[31] Urban agriculture supporters viewed the way the Manatawna project was handled as

> a prime example of the city completely failing a project. That wasn't the Office of Sustainability, but that was still a City thing, and they totally botched it from the beginning. The hard thing is, the wealthy neighbors that want to hold public land as their own backyard. They called us outsiders, they don't want any "outsiders" coming on public land in their neighborhood. They had money and influence and urban agriculture and the poor do not ... They [the City] did nothing to get public support for it, and they didn't work with Councilman Curtis Jones on it at all. Maybe 10 people wrote letters for it, and then he heard from 30 wealthy neighbors who were against it who then turned around and got petitions and falsely represented what was going to be happening out there ... But most people doing urban agriculture don't have time to go to Council meetings, or go get petitions or run really good campaigns. (Interview 12)

One interviewee suggested that the hopes that the city would develop significant opportunities for commercially viable urban farming were dashed with Manatawna: "a lot of the momentum kind of died when the big parks and recs project went down" (Interview 23). While the concept of urban farms and gardens might sound good in policy documents like *Greenworks*, the Manatawna RFI highlighted the fact that when it came to siting urban farms and gardens in communities, citizens and city councillors needed to be on board and included in the planning process. Parks and Recreation needed a different strategy for community engagement around supporting and promoting urban agriculture.

A New Approach: The Green 2015 Planning Process, Farm Philly, and the Community Farm and Resource Center at Bartram's Garden

The Manatawna RFI was viewed as a setback. Since then, Parks and Recreation has taken a more systematic approach to urban agriculture and worked with other stakeholders to develop citywide urban farming and gardening policy. The *Green 2015* planning process run by Parks and Recreation in 2010 in collaboration with other city agencies and non-profits solicited feedback from residents and various agencies and organizations regarding parks and green space.[32] The goal of *Green 2015* was to "transform 500 acres of empty or underused land into publicly accessible green space in neighborhoods throughout Philadelphia over the next five years."[33] In six "civic engagement sessions" held throughout the city, urban agriculture was identified as important to the city's greening plan.[34] After the *Green 2015* planning process, Parks and Recreation approached the city for funding to hire a staff member to coordinate the urban agriculture program.[35] The result is the Farm Philly program, which oversees more than forty farms and gardens located on Parks and Recreation land.[36] Through Farm Philly, Parks and Recreation is supporting urban agriculture on city-owned parkland and tying urban agriculture into its educational mission. Parks and Recreation's Junior Farmers program works with recreation centres and summer and after-school programs to teach students (ages two to twelve) about urban farming.[37] The department has also become a leader in the urban agriculture movement – it is helping to develop the Land Bank, participating in the city's Food Policy Advisory Council (FPAC), dedicating staff to promote urban agriculture in the city, and working collaboratively with other city agencies and the NGT to protect gardens and farms from development. An example of a successful project was in 2011 when Parks and Recreation partnered with the PHS, the University of Pennsylvania's Agatston Urban Nutrition Initiative (UNI), and Bartram's Garden to develop the Community Farm and Food Resource Center, a 3.5-acre farm and greenhouse, on the site of Bartram's Garden in Southwest Philadelphia. Residents of Southwest Philadelphia, a predominantly African American and low-income neighbourhood with limited access to fresh food, are offered opportunities to garden at the centre. The Community Farm and Resource Center provides sixty community garden beds where residents can grow their own food.[38] The centre also focuses on youth employment and education and has become a vital part of the community. In supporting projects like the centre, Parks and Recreation has taken a more community-driven approach to urban agriculture, making sure that the community is more engaged with the projects and is directly benefiting from them. This approach has been important to developing a local constituency that will support rather than fight against urban agriculture projects.

City Effort to Meet Greenworks' Goal of Promoting Commercial Farming: Marathon Farm

In 2011, in another effort to promote *Greenworks'* goal of supporting commercial farming, the city helped the Marathon Grill, a restaurant chain in Center City, secure both land and water for its urban farm (and a community garden for residents). The Marathon Farm in Brewerytown, in North Philadelphia, supplied fresh microgreens and vegetables to the Marathon Grill restaurants. City officials viewed the project as in line with *Greenworks'* mission of providing fresh, local food, and opportunities for community and youth engagement. It was also in step with the idea, popular at the time, of promoting economic development through urban agriculture.[39] Darrell Clarke, the city councillor for Brewerytown, was instrumental in helping the Marathon Farm secure land and gain access to water. However, some urban agriculture nonprofits questioned why the Marathon Grill owners had been given what they viewed as special treatment in terms of water access while other worthy programs – for instance, those that work with youth – were being asked to pay thousands of dollars for water access and were not being helped during the land acquisition phase. Interviewees were concerned that Clarke had essentially helped make the Marathon Farm happen, when – so they argued – he had not been as helpful to other farmers. Given that the owner of the Marathon Grill was also a large landholder in Philadelphia, questions arose as to why the city needed to help the Marathon Farm navigate the land acquisition process. The help extended to the Marathon Farm was viewed as further evidence of the problems that had long beset Philadelphia's urban politics: "Philadelphia's politics are so ingrained in the system ... Who you know? How much money you have? Favors? That usually doesn't involve the community" (Interview 17). However, city officials defended the Marathon Farm as an important project that would promote the connection of urban agriculture to sustainable business (Interviews 31, 26). The development of the Marathon Farm highlighted the tension around providing special treatment for some farms and gardens, the need for more transparent decision-making with regard to access to city property and water, and the need to value the community benefits of urban agriculture more systematically.

Ultimately, though, despite its initial success, the Marathon Farm did not work out as expected. By 2013 the Marathon Grill had closed many of its restaurants and pulled back its support for the farm, and the farm manager had left to work at the PHS (a trend among farm managers and urban growers; see below). With weeds growing all over the property, there was little sign that only a few years earlier this had been the site of a ribbon cutting with Mayor Nutter, as part of the "Greenest City in America" vision, which would have seen successful commercial farming all over the city.[40] Some of the neighbourhood residents who had been gardening on a section of the Marathon Farm site reserved for community gardens now mobilized to take over the Marathon Farm and develop the Brewerytown Community Garden. However, there was concern that some of the lots that composed the garden were being offered at sheriff sale. The community gardeners knew that the neighbourhood was gentrifying rapidly and that if they did not act quickly to protect the garden, it would be converted into condos.[41] Residents again lobbied Clarke for support, and he worked with Farm Philly to bring the land into Parks and Recreation's inventory, thereby preserving it.[42] Before Farm

Philly preserves a garden, it makes sure there is community interest in maintaining it over time; since the Brewerytown community gardeners were sufficiently mobilized, they were able to make a strong case for preserving the garden as a community space. The Marathon Farm to Brewerytown Community Garden story highlights the changing dynamics around urban agriculture in the city. Over the course of several years, vacant land could be turned into a commercial farm and community garden, become abandoned, become a community asset, come under threat of being redeveloped, and ultimately be protected by Parks and Recreation.

Allowing Growing in the City: The Zoning Code and Urban Agriculture

The first part of the chapter focused primarily on *Greenworks* and efforts by city agencies to promote specific farms and gardens. We now turn to citywide efforts to systematically include urban agriculture in policy-making. To meet the Food Charter and *Greenworks'* goals, the Office of Sustainability, through FPAC, a citizen and stakeholder advisory board staffed by the Office of Sustainability, in conjunction with numerous city agencies and non-profits, worked to develop policies to promote farms and gardens. Early advocacy around urban agriculture in Philadelphia mobilized around rezoning the city because urban farmers and growers recognized the rezoning process as an opportunity to embed urban agriculture in the city's legal code, following the lead of other cities. The comprehensive plan for the City of Philadelphia had been developed in the 1960s and by the late 2000s was extremely outdated. During the *Philadelphia 2035* process to develop a new comprehensive plan and zoning code, advocates for a green city made sure that urban agriculture and access to healthy food were included in the plan. *Philadelphia 2035*, the City's new comprehensive plan, adopted in June 2011, makes a number of clear statements about the importance of access to fresh, local food and legality of urban agriculture.[43] The plan aims to:

Provide convenient access to healthy food for all residents.

a) Maximize multimodal access to fresh food by encouraging grocery stores, healthy corner stores, and outdoor markets at key transit nodes and within transit-oriented development zones ...

b) Support agriculture and food distribution programs at recreation centers, schools, and other public facilities located in key neighborhood centers.

c) Establish farmers' markets along commercial corridors within neighborhood centers.

d) Increase local food production through zoning designations that permit urban agriculture as-of right in strategic locations and allow for roof-top gardening.

e) Develop standards and guidelines for community gardens and urban agriculture sites on public lands to ensure transparency, continuity of use, and community benefit.

f) Work with supermarket developers to create site designs that respond to neighborhood context and allow access for seniors, children, and other transit-dependent and mobility-limited populations.[44]

The new zoning code, passed in August 2012, allows "four subcategories of urban agriculture including animal husbandry, community garden, market or community-supported farm, and horticulture nurseries and greenhouses (see Zoning Code Chapter 14-601.11); permits urban agriculture and community gardens as-of-right in most residential and commercial districts; permits animal husbandry in most industrial districts; allows the square footage of fresh food markets to not count against the maximum buildable area for development projects."[45] Also, urban agriculture and farmers' markets are allowed on city-owned property (schools, parks, recreation centres).[46]

Proponents of urban agriculture were happy that urban agriculture was recognized in the city's comprehensive plan and new zoning code (and felt that it was only included because they had mobilized to have it included). But they were later dismayed when in January 2013, councillor Brian O'Neill proposed amendments to the zoning code in Bill 120917 that would compel urban farmers to apply for "special exceptions" on commercially zoned parcels.[47] A special exception would entail making a costly and time-consuming application to the Zoning Board of Appeals (ZBA). According to the GJLI, O'Neill's amendment would have put 20 per cent of the farms and gardens in the city of Philadelphia at risk and caused unnecessary paperwork, fees, and hardship.[48] Amy Laura Cahn, director of the GJLI, testified before City Council that O'Neill's amendment would undo the positive work that had been carried out during the comprehensive planning process. She argued that there were two problems with the amendment:

> One, the proposed amendments will have a negative impact on garden-
> ing and farming by jeopardizing existing, crucial community resources
> – many of which have been in existence for decades. Two, our current
> zoning code – the product of thoughtful deliberation, compromise, and
> public input – accurately reflects existing uses in commercial mixed-use
> districts. The code has made it simpler for gardeners and farmers to be
> in compliance, while providing standards to mitigate negative impacts.
> These zoning amendments build barriers back in.[49]

Ultimately, the amendment was defeated in 2013, in large part because of mobilization around the issue from Healthy Food, Green Spaces (which would serve as the foundation for Soil Generation), a coalition of groups including the GJLI, Weaver's Way Co-op, and the PHS.[50] Having to defend urban agriculture's inclusion in the zoning code reminded urban agriculture activists that they needed to stay engaged in policy-making to continue to make the case for urban agriculture to City Council and develop pathways to access and protect land for urban agriculture.

Urban Agriculture as a Temporary Use, but Not the "Highest and Best" Use

While the zoning now technically *allows* urban agriculture on vacant land in much of the city, urban growers recognize that legally *accessing* vacant land and protecting growing as a legitimate long-term use of the land is a different story. Even though urban agriculture has been formally recognized in *Philadelphia 2035* and the zoning code, there continues to be debate among city officials about whether it should be temporary or permanent. Urban agriculture does not pay the same kind of tax revenue that residential and commercial development does, which means that many city officials view it as temporary or transitional. A tension exists between urban agriculture activists, who see opportunities across the city for gardens and farms, and city officials, who are trying to balance the growing interest in urban agriculture with other city agendas and community needs. Urban gardeners and farmers are frustrated that the city has not taken a more active role in helping them access vacant land, particularly city-owned sites. Yet when asked about the difficulty of gaining access to vacant property for urban agriculture and the power that city councillors have to deny access through councilmanic prerogative, several representatives for the city explained that land access was much more complicated than the urban gardeners and farmers recognized (Interviews 24, 31, 26). One city official explained the "disconnect" between the urban farming and gardening view of urban land and the city's perspective:

> People feel that they should be able to do it [have access to vacant land for gardens and farms] since there is such a surplus of land … Figuring out what an appropriate system is that builds in flexibility and responds to some of the ideas that are coming up but also maintains not only the City's liability responsibility [the City is self-insured], but also the City's responsibility for community development. I always say to the farmers and gardeners our hope for the neighborhood that has many vacant lots is for them to be whole again. You wouldn't want to live on a block where only 1/3 of the houses are occupied. Neighbors want to see more people living there and more stores. Vacant land is part of that fabric right now, but may not be in the future. (Interview 24)

This city official was working to develop policies that balanced the needs of urban growers, the city, and the communities where the land is located. City officials recognize the need to make urban agriculture possible; at the same time, the city is planning other, more economically beneficial uses – specifically, redevelopment that will bring in tax revenue. The city officials we interviewed were very clear that urban agriculture was just *one* piece of the challenge of developing a more vibrant and sustainable city. Urban agriculture could play an important role, but it was one part of a larger vision of greening and redevelopment that included parks, streets, and economic development. Some city officials wanted to "wait and see" what happened with the economy before making long-term commitments to preserving urban space as farms and gardens. There is a clear tension between urban growers, who view gardens and farms as a long-term investment (particularly if they are going to spend the time to clean the lot and bring in water and plants), and city officials, who wonder, "How can you foster growing, as one of a number of interim uses, that can be applied to the city's vacant land at the appropriate scale?" (Interview 24). From the city's perspective, the "highest and best use" of urban land may be new development that expands the city's property tax base rather than an urban garden or farm, which may have significant community benefits and neighbourhood spillover effects but does not directly generate tax revenue. One non-government interviewee acknowledged the city's position and the need for urban agriculture activists to make a better social and economic case for urban agriculture: "The fact that the City is 30 per cent impoverished impacts its tax base so I do understand the City's interest in redeveloping land and we have to be able to make a greater case that these initiatives will help" (Interview 33). The city officials we interviewed were clear that they saw development as the key to the city's future. When asked whether Philadelphia was considering plans to promote urban agricultural districts, similar to those proposed in Detroit, one city official countered:

We are NOT DETROIT! We are growing. We lost 25 percent of our population. We were 2.1 [million] down to 1.5 [million]. Detroit was 2 million down to 700 with no bottom in sight. There is a really different dynamic. Our vacancy patterns do not lead you to believe that you can cut off an arm of the city and turn it into something else. Our vacant land is

> intimately interwoven into still operating real estate worth trillions of dollars. So we are not cutting off the development opportunities. Part of what makes cities attractive is density. Part of where we are today is to manage the blighting influence of land so it doesn't make density less attractive. Maybe there is a way to use interim cleaning and greening programs and community gardens as ways of making vacant land contribute to the quality of life in our neighborhoods and there is an opportunity to build back to a denser pattern. That is what we want to do because that is what a city is. (Interview 26)

City officials were working on strategies to "build back" the city by promoting development, and it was clear that in many cases, this, not urban agriculture, was a top priority. They recognized that urban agriculture could play an important (perhaps temporary) part in Philadelphia's resurgence, particularly at the neighbourhood scale. They knew there was excitement around urban agriculture, but the idea of converting whole parts of the city into farmland – a strategy that was being considered in Detroit – or even giving a large number of vacant lots over for permanent gardens and farms, was definitely not on the table.[51]

Developing Pathways to Access Vacant Land

Figuring out a strategy to manage the city's vacant land that allows for both redevelopment *and* access to land for community benefits (such as community gardens and urban farms) is an ongoing challenge. Ownership of vacant land in the city is extremely fragmented: of the estimated 40,000 vacant lots in the city, 14.3 per cent are owned by Public Property, 7 per cent by the RDA, 1.7 per cent by the Philadelphia Housing Development Corporation (PHDC), 0.1 per cent by other "city entities", and 76.9 per cent by "non-city entities" (or private landholders).[52] While there were calls by advocates of urban agriculture (and also by the development community) for the city to streamline the land disposition process, there was also distrust of the city, whose record of choosing development over urban agriculture (see the previous chapters) made activists hesitant to believe that Mayor Nutter's administration would take a different approach:

> People are frustrated right now ... There were more gardens in the 70s than there are now, and then the city came and developed them all. Kicked people off the land, and developed them, and gentrified a lot of neighborhoods. And that's a whole other crazy issue that could go on for hours. So a lot of people don't really trust the city ... They say a lot of things for green washing, and don't really back it up. They hire some great people, but don't let them do anything. (Interview 12)

Before the creation of the Land Bank (see below), the challenge for people interested in starting gardens on city-owned lots was that larger sites might be owned by more than one city agency (and/or private owner); this required the applicant to negotiate with each city agency separately. Urban growers who wanted legal access to a city-owned vacant lot first needed to identify the city agency (or agencies) that owned the land and then go through an extremely complicated and not very transparent land disposition process. If a garden or farm site encompassed lots that were owned by more than one city agency, the growers needed to work with *all* the land-holding agencies, a task that most urban growers, many of whom were volunteers, were not prepared to do. Urban growers who were interested in vacant city-owned land described navigating a maze of bureaucracy trying to get title to the land or permission to lease the land through the various city agencies. And for those urban growers who made it through the legal process of acquiring or leasing city-owned property, a final challenge was getting the councillor in the district where the land was located to approve the lease or land disposition. Since it is the councillors who make the final decisions about land use in their districts, urban growers found themselves at their mercy.

The process of creating a path to formal land access for urban agriculture has taken time and ongoing negotiation among stake-holders, who sometimes have competing objectives. In December 2011 the Philadelphia Redevelopment Authority's draft policies for urban agriculture alarmed urban agriculture activists, who felt that those policies had been developed without formal consultation with a diverse group of urban growers. The draft proposed urban garden agreements that could be broken at any time, as well as one-year leases and requirements for insurance and non-profit status for community gardens.[53] After activists in the Food ORganizing Collaborative (FORC), a group that "organizes food growers, food workers, and food eaters in the Philadelphia region to reclaim control of our food system and build food sovereignty,"[54] saw the draft version of the city's policies in December 2011, they quickly mobilized a series of community consultations (acknowledging that they did not have strong enough African American and immigrant representation at the meetings) to get feedback from as many farmers and gardeners as they could.[55] On 26 January 2012, FORC sent a letter to John Carpenter, deputy executive director, and Bridget Collins-Greenwald, deputy managing director, at the Philadelphia Redevelopment Authority (RDA), arguing for a city policy that better acknowledged the benefits and realities of urban agriculture. They argued that the proposal for

> Urban Garden Agreements – as temporary 1 year arrangements between the City and gardeners – will never provide the stability to allow garden projects to grow to their fullest potential – both financially and in their roles as anchor community institutions ... Gardeners should be able to easily secure leases on vacant land that are at least 3 years long with opportunities for renewal. Urban Garden Agreements meet the needs of few gardeners and should be the exception. These short-term agreements discourage gardeners from investing their time and labor into improving the land, as there is no guarantee of using the land a second year. At the very least these agreements should serve as an opportunity to demonstrate a garden's sustainability and a steppingstone to a longer-term commitment by both the gardener and the City.[56]

Urban agriculture proponents were also concerned that the draft's proposed requirements for non-profit status and insurance for community gardens, particularly new gardens, "may be prohibitively expensive and inaccessible at this stage."[57] The revised version of the city's land disposition policies addressed some of these concerns, and the city now offers the possibility of longer leases (and even purchase at reduced price) for community gardens, particularly those partnering with the Neighborhood Gardens Trust (discussed later).[58] Yet individual gardens are still viewed as *temporary*. The 2014 disposition policy states:

> The intent is to provide an interim use for the land to eliminate blight and improve safety until development is possible. The City expects that the majority of individual gardens will be temporary, and the land will still be available for development ... The Individual Garden Agreement (IGA) will last for one year, and may begin and end at any time throughout the year. IGAs are time-limited agreements that both the City and gardener may terminate at any time, with or without cause, upon prior written notice to the other party. The City will use reasonable efforts to avoid terminating the urban garden agreement between April 1 and November 1. The City will provide as much notice as possible if a license is to be terminated. "[59]

The one-year Individual Garden Agreements continue to reflect scepticism about the longevity of gardens. Some city officials were cautious about granting long-term leases to people who might quickly tire of the long hours, difficult working conditions, and seasonal low wages of urban agriculture. City officials also had stories of people using their lots in other ways. In one case, the site was being used as a makeshift car repair shop; in another, as an outdoor barbershop (Interviews 24, 31).

Community gardens and "community-managed open spaces" were granted more land security, with the possibility of longer leases (with city legislative approval), provided that they had insurance, support from a local organization, and a maintenance plan and showed evidence of community benefit.[60] Market and community-supported farms could be negotiated on a case-by-case basis with the city.[61] The city policy for formally gardening or farming vacant land was as follows: (1) individuals interested in gardening vacant land in the city were to submit an expression-of-interest (EOI) form and appropriate tax forms; (2) the councillor from the district where the land was located needed to approve the request – if it was approved, the city would issue a one-year individual gardening agreement (which the city could terminate at any time); (3) the applicant needed to have the proper insurance. Community gardens were to go through a similar process; however, the community gardens were eligible for longer lease terms of up to five years (with city legislative approval). This continues to be a high bar for more informally organized community gardens. Commercial farms needed to have a business feasibility plan and the proper insurance (leases of more than a year would have to go through additional city approvals and would take additional time).[62]

Coordinating Access to City-Owned Land through Philly Landworks

The development of Philly Landworks was another important step, one that made access to vacant city-owned land more transparent. In the summer of 2012 the Philadelphia Redevelopment Authority (PRA), in cooperation with other city agencies, launched Philly Landworks (originally called Front Door), a new online database of vacant land through which activists, residents, and developers could search for city-owned vacant lots and abandoned homes. Previously, to ascertain the owner of a vacant lot required checking with numerous agencies; now, with Philly Landworks, there is one Web portal, and this has streamlined the process of finding out who owns the land. But this tool, while making vacant land more visible to urban growers, has made that same land more accessible to developers. Some farmers and gardeners who were used to flying below the radar had mixed feelings about the portal, which formalized access to vacant land. Urban agriculture activists who "guerrilla garden" on vacant lots had concerns about the unintended effects of efforts to streamline the development process. They worried that the Philly Landworks portal might make it easier for land speculators to take over their gardens/farms. Once farmers/gardeners have invested in cleaning up vacant lots, land speculators might be interested in acquiring the land. Urban growers feared that developers might capitalize on the long hours spent cleaning a lot for an urban farm or garden by displacing the urban growers who had done the work. This is one of the ironies of policy reforms aimed at formalizing urban agriculture: these policies may be designed, in part, to support urban farms and gardens, but they also make it more likely that some informal farms and gardens will be targeted for displacement.

So even in the process of increasing access to land for growing, we see the underlying tension between the use value and the exchange value of land. Policies that make it easier to navigate the vacant land access process for urban agriculture do the same for commercial and residential developers, who tend to have much more expertise in land disposition and much greater access to financial resources.

One limitation of the Philly Landworks portal was that privately owned vacant properties were not included in the search. Another criticism was that while the city's vacant property databases had been merged, that database was not a land bank – that is, it did not clear land for sale or provide a clear acquisition process. Instead, Philly Landworks required prospective growers to submit an Expression of Interest (EOI) to acquire the land for lease or purchase. By February 2013 the RDA had received between forty and fifty applications for leases from farmers and gardeners, but the city faced a backlog of EOI requests, and it did not have enough staff to process them all.[63] In 2014 the Farm Philly program of the Department of Parks and Recreation was designated to act as the staff for the EOI process and to coordinate with various city agencies. This was an important policy development that underscored the city's interest in promoting urban agriculture, but growers were still frustrated by what they described as a stalled process. In the spring of 2016, the Land Bank (discussed below) took over the EOI process for city-owned land, asking everyone who had applied for an EOI for more information about their intentions to help them move through the next steps in the land lease and acquisition process.[64]

Getting Serious about Land Disposition: Developing the Land Bank

The need to develop and use vacant properties in Philadelphia has been recognized by residents, urban agriculture advocates, community development corporations (CDCs), and developers as well as by city officials who want to generate more tax revenue and redevelopment. A 2010 study found that Philadelphia's vacant properties represented $3.6 billion in lost wealth; the cost to each household in Philadelphia from lost property values was around $8,000.[65] The same report estimated that it was costing the city about $20 million a year to maintain the properties (mowing them and making them look clean and green); also, the city adds at least an additional $2 million in uncollected tax revenue every year on top of the $70 million in lost revenue from delinquent property (including interest and penalties).[66] Recognizing the costs imposed by vacant land and the opportunity costs of not redeveloping it, state and city elected officials, city officials, and activists (including urban growers) set out to develop a Land Bank that would streamline access to and redevelopment of vacant land.

Developing the Land Bank required policy changes at various levels of government. At the state level, HB1682, enabling legislation that allowed for the creation of state land banks, passed the State House and SB1414 passed the State Senate in April 2012. In Philadelphia in 2012, city councillors Maria Quiñones-Sánchez and Curtis Jones, Jr, proposed a city ordinance to create a city Land Bank Authority (Bill no. 120052); it was passed in 2013.[67] The Land Bank Authority, approved in 2015, is designed to streamline the acquisition of vacant land by establishing one authority with accountability, community representation, and transparency that will manage the city's vacant land disposition process. The Philadelphia Land Bank allows the city to acquire the properties, consolidate them, and offer them at reduced prices to groups interested in economic redevelopment as well as community development.[68]

The Philly Land Bank Alliance – an assemblage of community development groups, advocates of urban agriculture and local food, real estate agents, the building industry, and sustainable business professionals –advocated for a Land Bank in Philadelphia.[69] The Campaign to Take Back Vacant Land, a coalition of faith-based and community organizations, also called for a Land Bank, as well as for the inclusion of vacant land in community land trusts. One interviewee described the coalition that advocated for the Land Bank as "made up strongly of folks that care about affordable housing, folks that care about urban ag. and fresh food and accessing green. That group worked as hard as it could to get any of that in there. Additionally, some of those same people are actually on the Land Bank Board" (Interview 33). Proponents argued that the Land Bank could help promote the development of affordable housing and the public use of space, including for urban agriculture. They advocated borrowing the Land Bank and community land trust model from other cities, citing as an example the Dudley Street Neighborhood Initiative in Boston, where a community land trust developed affordable housing, stabilized the neighbourhood, and protected long-time, low- and moderate-income residents from being pushed out of their neighbourhood as property values rose.[70]

In Philadelphia, however, part of the tension around the Land Bank has been that it was designed to make it easier to acquire vacant land both for development and for community uses such as urban agriculture (primarily through a lease) – two uses often at odds with each other. While promoting the redevelopment of vacant land in Philadelphia remains the primary goal of many city officials, increasingly, city policies are formally acknowledging the value of urban agriculture as an intervening use (and in some cases a permanent use). FPAC, the GJLI, the HFGS (now Soil Generation), and many other urban agriculture proponents worked to ensure that urban agriculture was front and centre in the Land Bank disposition language and that the "use value" of urban land was recognized. Today the Land Bank recognizes the city's interest in urban agriculture: "The City supports the use of vacant land for urban agriculture that improves the quality of life in the City's neighborhoods."[71] Also, the Land Bank's disposition policies allow it to

offer properties at less than fair market value where the Land Bank finds that the proposed use would create beneficial community impact, such as affordable or mixed-income housing that is accessible or visitable; economic development that creates jobs for community residents; community facilities that provide needed services to residents; side and rear-yards; innovation in design and sustainability; urban agriculture; community open space; and any goals established under § 16-710. The Land Bank shall allow applications for less than fair market value, including nominal disposition, for any property owned by the Land Bank.[72]

In the 2017 *Land Bank Strategic Plan and Performance Report*, an exciting development is the possibility of landownership for community gardens that provide certain community benefits: "The City will consider selling publicly owned lots to garden organizations that commit to preserving and maintaining the land as a community garden or open space, particularly in areas with limited access to green space, fresh food and concentrations of low- and moderate-income households."[73]

The inclusion of farms and gardens as valued community uses eligible for below-market prices was a win for urban growers. While the Campaign to Take Back Vacant Land's vision of a citywide Land Bank that would connect to city community land trusts to promote affordable housing and other uses has not yet been fully realized, the Land Bank's contract with the Neighborhood Gardens Trust (NGT) is an effort to secure legal title (through ownership and leases) to the land for gardens and farms throughout the city. Working with the Land Bank, the NGT examined 328 gardens located across the city to better understand the conditions that make for good gardens so that they could develop criteria to review the current EOIs from people interested in leases for their community gardens.[74] In 2017 the Land Bank and the NGT plan to offer five-year leases to some of these approved community gardens.[75] Having the Land Bank partner with NGT is an interesting development, one that might be replicated in other cities. As mentioned earlier in this book, the NGA (the predecessor to the NGT) was founded in 1986 with help from gardeners, businesses, the PSU Gardening Program, and the PHS after some gardeners lost their gardens to development. The NGA was formed as a land trust that bought gardens' property rights in order to protect them from redevelopment. In 2012 the NGT created a partnership with the PHS, and this strengthened its ability to place gardens and farms in the land trust. Preserving gardens through a land trust is critical in Philadelphia, where there is increasing pressure for redevelopment as well as uncertainty around land tenure. As of 2017, the NGT has more than thirty gardens in the trust.[76] In the coming years, it will be interesting to watch how the NGT cooperates with the Land Bank and other city agencies such as Parks and Recreation and the Water Department to protect urban gardens and farms. In 2017, FPAC is discussing plans to develop a Philadelphia Urban Agriculture Strategic Plan, which may be an important policy tool "to guide departments in standardizing internal procedures and practices to support urban agriculture".[77] The city "will conduct comprehensive outreach to ensure the Plan reflects the vision, needs, and voices of Philadelphia's passionate gardening and farming community." [78] The willingness of city officials to develop a strategic plan is promising because it signals an ongoing commitment to urban agriculture and a recognition that the city needs to develop more comprehensive policies.

Is the Land Bank Too Late to Protect Many Gardens and Farms?

Notwithstanding the progressive goals outlined in the Land Bank strategic plan, activists continued to be frustrated by the slow launch of the Land Bank and the delayed response to EOIs for garden leases. Property redevelopment had picked up in Philadelphia, and once again urban farms and gardens were feeling the squeeze. An example of a farm fighting to stay in existence is the La Finquita Farm in South Kensington. With the help of the GJLI, the farmers at La Finquita are arguing for adverse possession (use of the land for more than twenty-one years) in order to stop the garden's demolition. Many of the activists who had been hoping the Land Bank could help them gain access to vacant properties were concerned that it had been established too late. According to the GJLI's Amy Laura Cahn (as quoted in *Philly.com*), "the garden [La Finquita] was waiting for the Land Bank to be ready to do just this. It had been pleading and reached out and started talking with the City Council president's office about using the Land Bank in exactly this way. But, like any number of other spaces, the Land Bank is too late."[79]

In Philadelphia's gentrifying neighbourhoods, protecting urban gardens and farms is a race against the developers, and garden demolition can happen quickly. At the Growing Home Garden in South Philadelphia, which is run by the Nationalities Service Center, an organization that supports refugee and immigrant communities, one day a bulldozer arrived and took the fence down on part of the site, which had been sold to developers.[80] Fortunately for the Growing Home Garden, it was selected as one of the gardens to be protected by the NGT.[81] Across the city, gardens and farms continue to be displaced, but since many gardens and farms have been operating informally, it is difficult to estimate how many gardens and farms have been lost to new development.

For many, losing a garden or farm is a genuine community loss and can be traumatic. In September 2016, at the City Council's hearings to develop citywide urban agriculture policy, urban growers talked about how they had lost their gardens and farms or were at risk of losing them. Jamillah Meekins of the Master's Work Community Garden in Brewerytown/Sharswood testified about the history of the garden and the community's unsuccessful struggle to protect it from sheriff sale. Her story highlights the challenge of protecting urban gardens and farms from rapid gentrification and redevelopment in previously disinvested minority communities:

My grandmother came to Philadelphia from the South in the 1940s and settled in the Sharswood community where she worked, raised my mother, my aunts, and my uncles. She grew food for her family on three plots of vacant land on our block for many years. As the years passed and my grandmother was no longer able to work in the garden, the land became overgrown with weeds and trash. I watched as the entire community became neglected and infested with urban blight. My neighbors and family decided to work together to turn the Master's Work Community Garden back into the community food source it once was. We worked diligently to clear the weeds and trash to set up planting beds in their place. Several neighbors began to grow food to supplement the nutritional needs of their families. Historically, there's been no neighborhood supermarket available to the Sharswood community. However, there's a plethora of corner stores that peddle unhealthy foods to our children. The Master's Work Community Garden was our answer to the lack of food access. We were aware that two plots of the land that we grew on were owned by absentee landholders and the other was owned by the City of Philadelphia. We made attempts to request legal use of the land through the EOI process, but there was no success. One day we received notice that one of the plots of land was going up for Sheriff Sale. Once again, our neighbors pulled together and gathered their financial resources to make a bid on the gardens lot. Unfortunately, we were outbid at the Sheriff's auction. Our neighborhood has seen an influx of new developers and individuals with interests in making a profit from the blight that has been allowed to exist in our community."[82]

Like the Master's Work Community Garden, other gardens and farms across the city are meeting a similar fate, many of them silently. The GJLI website posted a photo of a note left by a gardener on 19th and Lombard, an area that is rapidly developing, that read: "I have loved this garden every day over the past 4 years. Please help yourself to the organic green beans, wildflowers, herbs, pots and soil, etc. before a builder destroys it completely on the 30th."[83] While this gardener with more resources was able to alert the GJLI about the garden's loss, silent displacement may be especially common among small, shoestring farms and gardens being run informally in disadvantaged neighbourhoods by local residents, who may not be alerting the news media or city councillors when they are threatened with displacement.

While the Land Bank can be an important tool for protecting gardens, it also has its limitations. Urban growing activists see the "Land Bank being passed [as] a huge thing with awesome potential opportunities for folks to preserve land" (Interview 32). But they also caution that it is

> still a really slow, complicated, bureaucratic, time consuming process to go through the Land Bank and preserve land and that process disproportionately advantages people who have experience which are usually not smaller grassroots projects. We literally did not have the social capacity to navigate the Land Bank. Collectives of college educated young people can probably make it through the process. It advantages this wave of outsiders. The process is imbalanced. (Interview 32)

Even in 2017, stakeholders from the city and urban agriculture activists are still figuring out how the Land Bank will operate. The organization has a new board of directors (with membership from CDCs), a new strategic plan, and an official role in the city. However, there is also some question about whether the Land Bank's recommendations are being recognized by other city departments, particularly the Revenue Department, which can put land up for sheriff sale.[84] For one garden activist, participating in the Land Bank strategic planning process was eye-opening because of the city's continued focus on the "highest and best use" of land and the need to frame urban agriculture in terms of the economic, community, and environmental benefits it provides (Interview 36).

The Challenge of Formalization: When Urban Agriculture Is No Longer "under the Radar"

While there is general agreement that streamlining and formalizing urban agriculture is a means of supporting it, there is also fear about what formalization will mean for farms and gardens that have long operated "under the radar." In this new paradigm, city officials will ultimately have to make tough decisions about how to handle the network of "unregulated" or "guerrilla" gardens across the city. Sometimes these gardens are located on private land where the gardeners have made agreements with the owners. In other cases, activists have taken over vacant lots without permission. As the city formalizes urban agriculture and even asks gardeners and farmers to secure insurance and leases for city-owned land, lots for growing – which once existed under the radar and were ignored – may become problematic. One city official expressed concern about gardening and farming on city-owned property: "Now I have to care because people do things on our property and we are liable" (Interview 26). Since the city does not have insurance and self-insures, this makes Philadelphia relatively risk adverse. The question of insurance is still being worked out.

To really promote and protect urban agriculture, a whole system of regulation will need to be developed. The formalization process will also require identifying the current agricultural activities in the city, and this means knowing what the current land uses are. Since many urban growers are used to operating under the radar, it can be difficult to make them visible. The GJLI's Grounded in Philly project uses Web-based GIS to show the locations and uses of vacant land. By crowdsourcing information about where people are growing food on vacant lots, Grounded in Philly helps make visible urban agriculture projects throughout the city. Urban agriculture activists hope that once they can show just how many gardens and farms the city has, they will have more power when it comes to lobbying for their protection. But because Grounded in Philly relies on crowdsourcing to account for gardens and farms, it is difficult to have an up-to-date, accurate count of active gardens. As of this writing, many city policies around urban agriculture are still being formulated and there are no formal city-led efforts to kick people off land. In one case, the city came by to check on the "state of a garden," saw that it was not well kept, and did not renew the garden's lease.[85] In a petition to Councillor Kenyatta Johnson, this gardener talked about how the city had judged the state of the garden and decided not to renew the lease:

> As a result of our temporary negligence, or perhaps because of growing pressure from developers who would buy the garden at a great profit to the city, the City has decided not to renew the individual garden lease for 2016. This is a devastating blow to a project that, in under a year, completely transformed a vacant lot into a productive growing space.

In fact, the garden is large and productive enough to provide produce
for several city families for 5-8 months per year.[86]

This note is telling because of the gardener's scepticism about
the city's intentions – it hints that the garden was denied a lease
because the city supported redeveloping the lot in a gentrifying
neighbourhood, a practice that is likely taking place in neighbourhoods across the city.

Soil Safety Policies

Another challenge that comes with the formalization of urban agriculture is soil safety. As a part of FPAC, a Soil Safety Working
Group was established in the summer of 2014 that brought together city and state officials, academics, public health advocates,
urban growers, activists, and community development groups.[87]
This group developed protocols for how urban soils should be
tested and monitored, as well as best practices for urban agriculture (particularly using raised beds). The policy recommendations
are mainly voluntary and are modelled on the soil safety protocols
developed in other cities. They represent the first step in a longer
policy dialogue about how to promote safe urban agriculture. The
challenge is to develop policies that take into account the realities
of urban farming (acknowledging the lack of access to financial
resources) and that keep people safe without erecting yet more
hurdles for urban growers to jump over. As a result of the Soil
Safety Working Group's work and initiative by city officials, the
city secured a $200,000 Environmental Protection Agency grant for
brownfield assessment on twenty city-owned urban gardens and
farms.[88] Sarah Wu, deputy director of the Office of Sustainability,
was quoted in *Plan Philly* explaining the importance of the EPA
grant to address issues of soil safety:

We kept getting questions about soil safety, whether it was safe to grow
in the soil in Philadelphia, and we didn't feel that we had the answers
… Now we can know what soil we're growing in. So if you find that

there's no contamination in it, you can erase your fears. And if you find contamination, there is EPA money available to apply for remediation. So basically, if we find a problem we can fix it.[89]

The safety of growing in urban soils certainly needs more examination in Philadelphia and other cities. City officials and growers alike are interested in this question. The ability to connect to other resources and expertise about brownfield remediation through the EPA is an important step towards more systematically addressing concerns.

Accessing Water and the Stormwater Exemption Fee for Gardens

Urban agriculture activists expected there would be a natural synergy between the goals of the Philadelphia Water Department's *Green City, Clean Waters* plan and promoting urban agriculture across the city. Despite PWD's early leadership in urban agriculture, this has not yet been the case. In fact, access to water continues to strongly impede the development of more urban farms and gardens across the city. The prohibitive cost of water hookups prevents many urban farms and gardens from having direct access to water. The PWD has posted a guide on its website explaining how gardeners and farmers can access water for their gardens and farms. However, it reads like a legal and engineering manual, and the hookups are cost prohibitive for many farmers and volunteer gardeners who do this as a hobby, whose gardens are non-profits, and who in most cases are not breaking even selling their produce. The PWD acknowledges the challenges facing urban growers who want access to water and who may not be able to pay hookup fees or cover the city's stormwater fees. The factsheet states that the PWD needs to work with the PHS and gardeners to identify ways to grant water access. Also, the city offers a no-interest HELP loan called the Urban Garden and Farm Loan to approved gardeners to help them pay for water hookups.[90] This is a step towards making water more accessible for urban growers, who still, however, are required to make significant long-term investments to access water (often on land to which they have no formal rights). Since most gardens and farms operate on shoestring budgets and rely on volunteers, the barrier to formally accessing water is still there, and this has left many gardeners and farmers relying on collecting rainwater, borrowing water from neighbouring houses, and using fire hydrants.

A critical tension between community gardeners and the PWD was the stormwater fees charged on urban gardens. The PWD adopted those fees in *Green City, Clean Waters* to encourage land-owners to install green infrastructure on their property in place of impervious surfaces. The gardeners felt they were being treated unfairly because even though they were effectively providing green infrastructure through their gardening efforts, the PWD considered the land to be impervious and charged them the same fee that a parking lot would pay. In June 2016, gardeners testified to City Council in support of Bill 160523, sponsored by Councillor Quiñones-Sánchez; the bill would provide a stormwater fee exemption for community gardens, in recognition that keeping the land open for gardens would provide valuable ecological services and act as green infrastructure – the exact goal the PWD was trying to achieve through *Green City, Clean Waters*.[91] In sponsoring the bill, Quiñones-Sánchez argued that there was a synergy between urban agriculture and stormwater management: "Exempting gardens from stormwater fees is a no-brainer because they naturally absorb stormwater run-off that would otherwise burden our sewer system. Furthermore, developing community gardens is critical to our strategy to repurpose vacant and blighted land into green spaces that our neighbors can share."[92] The passage of the stormwater exemption fee in 2016 was an important win for advocates of community gardening. However, one grower expressed frustration at how difficult it had been to get this concession from the city: "It was a fight to get the stormwater exemption passed because gardening is not valued" (Interview 36).

The change in the stormwater fee policy is further evidence that city agencies and city councillors are recognizing the need to support urban agriculture. However, as is the case with most urban agriculture policies in the city, change primarily happens as a result of mobilization by urban agriculture activists, who consistently argue that urban agriculture is a legitimate land use that is worthy of city support.

Urban Agriculture and Urban Sustainability: Developing the Framework

In Philadelphia the policy framework surrounding urban agriculture was described by one of our interviewees as "'hopeful' ... I think there are more groups than ever that are focused on these issues and focused on making policy change that will allow equitable access to fresh, healthy, and affordable food" (Interview 33). A number of institutionalized efforts are supporting urban agriculture, and these may provide lessons for other cities. City agencies have experimented with promoting urban farms. The Philadelphia Food Charter and *Greenworks* have made urban agriculture a goal. The Office of Sustainability is now permanent (having been approved by citywide vote to change the city's charter), and FPAC is continuing to evolve and expand its participation in governance.[93] The Land Bank Authority has been created. The Farm Philly program under the Department of Parks and Recreation is off and running, and the new comprehensive plan and zoning code acknowledge urban agriculture. Groups like the NGT and the PHS continue to be there to provide financial and capacity-building support to urban growers, partnering with city agencies. In addition, a number of citizens' groups and activist groups like the GJLI serve as watchdogs promoting urban agriculture. Coalitions like Soil Generation are increasingly pushing the city to acknowledge the equity implications of urban agriculture, particularly around race and class dynamics.

At the city level, agencies are recognizing urban agriculture as an acceptable (and in some cases even desired) use in their official documents. Efforts are being made to protect and promote urban agriculture through more formal channels; however, these efforts are slower and less comprehensive than urban agriculture activists would like. There is also a distinct difference between what is written in city documents and what happens on the ground, where councilmanic prerogative still acts as a barrier. Philadelphia has made progress in developing a system to streamline access to vacant land, developing gardening and farming leases and protocols, rezoning to allow for urban agriculture, and developing a Land Bank to promote efficient disposition. However, the formal process of submitting an EOI for vacant land to be used for farms and gardens is still relatively uncertain and complicated, and the terms of the leases are still perceived as too short for gardeners and farmers. There continues to be a lot of distrust between growers and the city, and some of the former prefer to "fly under the radar" (Interview 37). In addition, important administrative questions are being worked out, such as how the Land Bank will operate in practice, how urban agriculture will be prioritized as a land use, and which agencies will provide staff to implement the various new policies.

The pieces of urban policy for urban agriculture are there with the new zoning code, Farm Philly, the NGT, the Land Bank, FPAC, and groups like Soil Generation, the GJLI, and the PHS. However, interviewees thought that political will in the city government was still missing:

> The urban agriculture movement is growing stronger. A lot of younger people are getting more connected and collaborating and building networks and thinking critically about their work. I think the City needs to take a stand that supports this movement ... Aside from our community of urban ag. advocates, in many other circles in the City it quickly falls off the priority list. (Interview 36)

A positive development for urban agriculture is that in May 2016 the City Council approved a resolution calling for public hearings on urban agriculture, to be held by the council's Committee on the Environment.[94] The resolution recognized the progress that had been made in Philadelphia around urban agriculture as well as the need to support the "growth of urban agriculture in the city of Philadelphia."[95] That city councillors were taking urban agriculture seriously was encouraging, but there were concerns that City Council's renewed interest in urban agriculture was primarily around the possibility of developing vertical, commercial farms (such as Metropolis Farms in South Philadelphia, which grows crops vertically in a warehouse) and might not be as supportive of community gardens and farms that do not provide the same types of economic benefits (Interview 32, 36, 37). In calling for the hearing, one city councillor made it clear that the economic potential of vertical was his *primary* concern. Councilman Al Taubenberger was quoted in a *Newsworks* article: "The goal of these hearings is to establish the city of Philadelphia as one of the world's most prominent training centers for this type of [vertical] farming … The second call is for the expansion of urban farming in the city … It's very important for families, for nonprofits, producing fresh vegetables right here in the city of Philadelphia."[96] In September 2016, when the hearings were called, Soil Generation and the GJLI mobilized a diverse group of around twenty urban agriculture supporters to testify about the social, economic, and community benefits of urban agriculture. They asked the city to recognize these benefits and provide more formal support.[97] The City's interest in developing a Philadelphia Urban Agriculture Strategic Plan is also promising, indicating a growing City acknowledgement of the value of urban agriculture and the need to develop citywide policies.[98]

Policy-making around urban agriculture in Philadelphia, as in other cities, is still in many ways in its infancy. However, the more urban agriculture can be recognized as an important and valued component of a healthy, vibrant, and sustainable city, the more city councillors, the mayor, and city officials will work to institutionalize it. One interviewee highlighted the importance of connecting urban agriculture to the city's other important redevelopment goals:

> I do think that in a city that has 30 per cent poverty, the poorest big city in America, the urban ag. community has to be able to make the case to our leaders that this is a use that is not just good in theory, but that can put people back to work and help those folks within communities that are impoverished. And I don't know that we are there yet. We want to be there. It is well intentioned, but I haven't seen the complete merging of those worlds. And looking at the urban ag. world, it is largely

white and even though it is well intentioned … We have not arrived yet. (Interview 33)

In a city facing mounting pressure for real estate redevelopment combined with real economic challenges and racial and economic inequity, making the case for urban agriculture will require urban growers to continue to engage politically, particularly with city councilors, as well as a real commitment on the city's part. The hearings at City Council were an important step in this ongoing policy conversation. Amy Laura Cahn from the GJLI wrote in a May 2016 *Grid Magazine* article that unless the new mayoral leadership in Philadelphia quickly commits to urban agriculture as a permanent use, the gardens and farms will lose out to developers:

> In a decision between a decades-old garden or a new development, we do not know what will win out.
>
> If these decisions are made on a case-by-case basis, gardens will get picked off one by one. We need the Kenney administration to direct all relevant agencies to work together on a coordinated approach to aggressively protect the investments that communities have already made. And we need this city to retire the notion altogether that gardens are simply interim use. Generations of Philadelphia gardeners have proven otherwise.[99]

In Philadelphia, we anticipate that the conversation about the need for more permanence for urban agriculture will be ongoing, as it is in other cities. If anything, as development pressure picks up, the conversation may become more difficult as gardens and farms come into direct competition with developers for "valuable" land. One grower wondered about the future of urban agriculture: "I think the biggest challenge is making sure that land that people are growing on is not over built because there is literally construction everywhere. It is not just my neighborhood. It is South Philly. It is West Philly. It is everywhere" (Interview 35). Another interviewee was concerned with the slow pace of protection and emphasized the need for the city to act quickly: "These gardens are really hanging in the balance because there isn't much security and there isn't any real path to security" (Interview 36).

chapter five

"New Growers" Making Sense of Their Role

"It is kind of like the art movement of the 60s. All of the sudden young people are coming to Philly to do farming. What the city has is that you can still stay current to what is going on, but you can also farm. For a certain niche of people, it has really good value."

Interview 25, Speaker 2

As noted in the earlier chapters, Philadelphia has a long and rich history of community gardening and urban agriculture. In the 2000s, growing in the city became increasingly popular among young, often white, college-educated professionals with little to no experience digging in the dirt. This chapter explores the current interest in urban agriculture through interviews with people who consider themselves part of a "new wave" of urban growers, many of whom are young, white, and college-educated. It is important to understand these young growers because they do much to frame how urban agriculture is viewed in Philadelphia. They help shape the conversation about urban agriculture in the city and are engaged in policy-making and place-making. At the neighbourhood level, they are working to transform spaces, often in communities that are not their own and/or have long been disinvested, low-income, and primarily African American. Meenar and Hoover's 2012 study found a "high percentage of White gardeners in some predominantly non-White neighborhoods."[1] This raises important questions about the role and impact of these "new growers." They are lobbying for more access to land and ensuring that urban agriculture stays on the city's agenda. They are also helping define the role of urban agriculture (for better or worse) in the city through their interactions with communities and their connections with more established growers. Since 2011, when we started this research, we have seen a change in this "new wave of growers": many of them have been joining forces with older growers and people of colour, who argue that issues such as gentrification and control over land need to be at the front of conversations about the impact of urban agriculture on the city. Groups like Soil Generation, "a black-led, grassroots coalition of radical community gardeners and urban farmers," have been working with white farmers to help them understand their role, positive and negative, in low-income communities of colour.[2]

While there are many older white, African American, Puerto Rican, Southeast Asian, and other immigrant gardeners, and more and more younger activist African American farmers and growers, we focus in this chapter on the perceived "new wave of urban growers" because we want to simultaneously problematize and understand how they are interacting with the existing urban growers and with the communities where they have chosen to garden and farm. There is common ground between the new generation of growers and the older growers with regard to the importance of growing in the city; that said, the two groups' motivations are somewhat different.

The new growers offer a multitude of reasons why they see urban agriculture as important: some want to grow because they want to be more connected to nature; some are primarily interested in fresh food; some are drawn to the local food movement; and some see urban agriculture as necessary for Philadelphia, which is facing a host of challenges. Some are focused on social justice, are very conscious that their work involves place-making, and are highly aware that they are outsiders in disinvested, disenfranchised, often African American communities. Others are less troubled by race and class tensions and are intensely proud of what they have been able to accomplish: they see the fact that land goes from being blighted to being productive as a success. Their reasons why they get into urban agriculture are complicated and nuanced, just like everyone else's reasons for doing anything. In this chapter we explore how this new generation of urban growers makes sense of their role. We realize there is a diversity in their views that we cannot fully capture, but in examining these new growers, we hope to take a step back and ask questions about the part they play. Recognizing that this is a limited study, we also call for additional research that focuses on race, class, power, and community and how the dynamics of these relate to urban agriculture and urban change.

"A Really Excited Moment": Finding Meaning through Urban Agriculture

Not surprisingly, an appetite for local and fresh food served as the gateway for many growers' interest in urban agriculture, in Philadelphia and beyond. At the national level, a growing awareness of the environmental impacts of industrial agriculture has been complimented by efforts like Michelle Obama's White House organic garden and the Farm to School initiative, both of which highlight the importance of local fresh food, particularly in low-income communities.[3] Local and sustainable food has long been popular in Philadelphia, which is home to the White Dog Café, a nationally known restaurant specializing in sustainably sourced foods started by Judy Wicks; Greensgrow Farm, a profitable urban farm started by the late Mary Corboy on a former brownfield site in the Kensington neighbourhood; and the Weaver's Way Co-op, a food cooperative that has been hugely successful and that in 2013 generated an estimated $10 million in revenue.[4] As noted in previous chapters, the city has a long history of urban agriculture and a strong non-profit sector that supports urban agriculture and greening. Sometimes called "Farmadelphia" because of the possibility of reusing its vacant land, Philadelphia is also well situated geographically, close to some of the most productive agricultural land in the country.[5] Its farmers' markets offer fresh local food from the city and the surrounding communities.[6] Urban agriculture as an alternative use of urban land in shrinking post-industrial cities has gained the most national attention in Detroit, where until recently, with Quicken Loans' investment, in many parts of the city there was little demand for redevelopment.[7] Philadelphia's situation is different: it is wrestling with deep and persistent poverty even while many neighbourhoods are rapidly gentrifying.

While urban agriculture across the United States and internationally is a growing trend, the personal stories of young, primarily white growers suggest that their interests in urban agriculture are complex. For these young, college-educated, altruistic, underemployed, and sometimes frustrated growers, farming is more than a hobby or way to put fresh tomatoes on the table; they view urban agriculture as offering solutions to many of the social, environmental, and political challenges that young adults perceive in today's world, although many of them struggle to explain precisely how. When asked why they decided to farm in the city, interviewees talked about farming's links to social justice, anti-racism, anti-globalization, environmentalism, and politics. One urban farmer explained:

> It's this really excited moment. It's definitely tied into the whole DIY return in general. I think it's about alienation kind of ... It definitely has to do with the green frenzy, but it's kind of a spin-off. I started as an environmentalist, and left that and went to farming because I thought it was like more integrated, kind of, I felt like I could touch more issues with farming. I was interested in anti-globalization, I was interested in anti-racism, and I was interested in, whatever, and farming. Environmentalism was kind of like, the food touched all those things. I think it's about that, kind of, but now it's become this whole other aspect. This weird foodie culture that's getting really huge. Everyone is really into fancy restaurants and eating local produce. And I think that's not really about the politics, just about taste. I don't know, maybe it's about the politics a little bit. I don't really know [laughs], I think about this, and it's hard for me to wrap my mind around what's cool about farming right now. (Interview 15)

This justification for urban agriculture was often voiced by the young (in their twenties and thirties) urban farmers we interviewed.

They would connect multiple goals to urban agriculture, which for them seemed to be a piece of a much larger story, almost a panacea. They could explain their involvement in urban agriculture primarily by linking it to other values that were important to them. The reasons why people are drawn to urban farming are often personal, but at the same time, there are common forces behind the desire to grow food in the city: the local food movement, concern about environmental and social justice, a resurgence of interest in the city, a do it-yourself (DIY) mentality, and, in economic downturns, the limited economic prospects faced by recent college and university graduates.[8]

Urban agriculture also responds to Philadelphia's current urban condition. The city's high levels of poverty, food insecurity, and vacancy (which many of our interviewees mentioned as key to their interest in urban agriculture) align with its rich history of greening. Urban growers are taking advantage of the city's strong institutional base supporting their work. As the earlier chapters made clear, urban agriculture is not at all new in Philadelphia; however, many members of this grassroots movement of young people see themselves as innovative, as part of a larger movement, many of whose members are white and college-educated and want a new direction for cities. For many young growers, urban agriculture is central to their professional and personal identity. Those who are busy clearing vacant lots, growing food, and mobilizing around social justice and sustainability see themselves as a part of a growing movement of young people who are rethinking how cities should function. Some of the interviewees pointed to the work of Richard Florida and even referred to themselves as the "creative class" (Florida 2014; Meenar and Hoover 2012). According to Florida, the defining characteristic of the creative class is that "its members engage in work whose function is to 'create meaningful new forms'" (Florida 2014, 38). Originally writing in 2002, Florida probably did not envision that a group of unemployed and underemployed college graduates in their twenties (in September 2012, Philadelphia's unemployment rate was 10.9 per cent)[9] farming vacant lots would see themselves as a "creative class" that was revolutionizing the food and economic system, but this is a part of the narrative they tell.[10]

Some of the interviewees had moved to Philadelphia specifically because of its urban agriculture scene. One gardener described Philadelphia as offering unique opportunities in urban greening:

> I came to Philadelphia for a gardening job. I was really excited about it. The network is here and you would really want to start making connections, the opportunities to get involved snowballed and there is something to do garden related or horticulture related almost every weekend starting in April through October with all the different horticultural societies and arboretum and different activities that are happening in community gardens. (Interview 7, Speaker 2).

That several urban gardeners and farmers we interviewed moved to Philadelphia with the intention of getting involved in urban agriculture was especially notable in the context of Philadelphia's dramatic and persistent population loss over the past sixty years. Although urban agriculture does not currently provide many jobs and will not likely be a major employment engine, the presence of Philadelphia's strong urban greening network is attractive for millennials who are looking for opportunities to garden and farm and to be a part of a larger sustainability movement. Some are attracted to Philadelphia for this reason; for others, it is a reason to stay in Philadelphia after college.[11]

In fact, in urban farming circles, Philadelphia is viewed as having an edge over other cities. Every month, *Grid Magazine*, Philadelphia's urban sustainability magazine, focuses on some aspect of urban agriculture in Philadelphia. Often the cover shows a group of young people, sometimes but not always white, looking proud and growing food in what was formerly a vacant lot. With its photographs of lush vegetables and smiling faces, the magazine highlights the importance of urban farming in the city. Philadelphia's large swathes of urban disinvestment and perceived abandonment serve as a magnet for young people who want to grow in the city. Philadelphia has the amenities of a major city, but it also has land and an urban agriculture movement to join:

> When I moved to Philadelphia it was like, why are you going to Philadelphia, what's in Philadelphia, and now it's like people know

about it and want to come here and I've talked to a lot of people who are younger and they are like "yea, we want to move to Philadelphia," so I think their energy is the driving force. (Interview 7, Speaker 1)

We do not know how many people have moved to Philadelphia specifically for urban agriculture (and the number is probably not very large), but we do know that the interviewees talked about the unique opportunities Philadelphia offers for young people who are searching for non-traditional careers in urban agriculture. According to the 2010 Census, Philadelphia is a popular city for young graduates; its youth population (20–34) has grown by 14.7 per cent since 2000.[12] Part of this may be explained by economics: the cost of living in Philadelphia is significantly less than in other east coast cities, although housing costs are rapidly increasing.[13] Many of the young urban growers whom we spoke to talked about the importance of Philadelphia's affordability, particularly since urban agriculture provides low-wage and primarily seasonal work. They needed to be in a city where they could piece together an income that would cover their rent and living expenses. Of course, this was not the only factor about Philadelphia that appealed to these young growers, but it was certainly an important consideration. As the cost of living increases in Philadelphia, it will be interesting to see how sustainable urban agriculture is as a profession. Since 2011 we have seen some of the farmers and growers who initially were piecing together small incomes so that they could be involved with urban agriculture, shift gears and join Philadelphia's network of professional greeners, who work for the city, non-profits, and universities. Of course, the transition to more traditional professional jobs is also a result of the economy picking up, and urban growers recognize that urban agriculture alone is not likely to offer them long-term financial security. Many interviewees who worked on urban farms and gardens hoped one day to support themselves on urban agriculture alone; but many of them also recognized that steady employment with one of the many non-profits in the city focusing on greening and food could allow them to stay true to their mission and earn a living, while they cleared vacant lots and "guerrilla gardened" (which some admitted was their true passion) on the side.

As the first chapters of this book highlight, key to Philadelphia's success as a hotbed of urban agriculture is its long history of institutionally supported urban greening and its long tradition of growing in the city. To start their farms and gardens, urban growers networked at PHS workshops and looked to that society for advice on how to start growing in the city. Many of the interviewees participated in the PHS's Garden Tenders education program, which teaches participants how to start a garden and connects them with a network of other people interested in community gardening. According to a PHS employee, the society started Garden Tenders because "we couldn't keep up the volume of requests for people who wanted help to start gardens." One community gardener explained the importance of PHS to growers:

> It's true, a lot of people, almost anybody that's running an organization right now either worked for or came from a program of PHS. They are the nexus of urban greening in the city. If we didn't have them, I don't know what we would do. (Interview 11)

Besides the Garden Tenders program, the PHS provides training and gardening supplies for members of the City Harvest Growers Alliance, which in 2015 was a network of 140 farms and gardens in the city.[14] A portion of the food grown at the CHGA gardens is donated to food cupboards through the City Harvest Program; this directly links the urban gardens and farms with food security in low-income communities.

In addition to traditional greening organizations like the PHS, there are networks like the Philadelphia Urban Farm Network listserv (started by local farming activists) that help urban agriculture activists and enthusiasts connect to other city and non-profit resources and share experiences and concerns. PUFN members[15] often help one another navigate the challenges they face starting an urban garden or farm.[16] Groups like the Garden Justice Legal Initiative, a legal advocacy group, help gardeners and farmers chop through the thicket of bureaucracy that surrounds land access and tenure. Legal support from GJLI has helped save a number of gardens across the city from residential or commercial redevelopment, and GJLI has sometimes helped citizens purchase gardens and farms and/or put the land in the Neighborhood Gardens Trust (NGT). These institutional resources have been critical to promoting and supporting urban agriculture in the city.

Urban Agriculture: An Identity and a Career?

The growers we interviewed indicated that urban agriculture played an important role in their lives. When asked why she was making a career in urban agriculture, one interviewee said:

> I think it is something a lot deeper ... I think it is the basic level of feeling very strongly about not being connected whether it is to nature, being outside, a little bit like that nature deficit disorder. There is also new knowledge. Or the recent food scares. People realize, I don't know where my food comes from. That is coupled with the economy. It is a lot of different things that ha[ve] all come together at the same time. (Interview 25)

Urban agriculture offers an alternative for socially progressive young adults seeking their identities; it also offers a niche in an economy they perceive as racked by corporate dysfunction and growing social and economic inequalities. For millennials, urban growing symbolizes positive change in an era when many of them are frustrated with the current state of our cities, our food system, and the jobs (or lack of jobs) available to them. Increasingly disillusioned with traditional job pipelines, many young college graduates are struggling to find meaningful and rewarding work that matches their interests, especially during times of economic recession. Urban agriculture is framed as an alternative. Many of the young farmers we interviewed were able to make ends meet through part-time jobs, such as babysitting and waiting tables, while pouring most their time into new farming and gardening projects. These new urban agriculture ventures may seem risky – what is the likelihood that a person will make a living selling collard greens in a challenging urban environment? – yet some of the growers are very clear that they really do not have much to lose:

"in an economic decline, people have a lot of time on their hands since a lot of people are becoming unemployed, so a lot of people are willing to try new things" (Interview 11). This connection between unemployment and urban agriculture is reminiscent of the vacant lot cultivation efforts in Philadelphia from the early 1900s, when in times of hardship unemployed workers were encouraged to grow food on vacant lots to feed their families and keep up their own morale. However, the consciously anti-corporate and political role of urban agriculture today is different and perhaps harkens back to the activism of the 1970s. There is a growing disgust with the food system and a desire to bring about change, perhaps one garden plot at a time.

But in contrast to the early 1900s, urban agriculture today is not just for the unemployed and the hungry; it has become a new career goal for some young professionals and a part of their collective identity. However, jobs in urban agriculture are not easy to come by, nor are they easy. Becoming an urban farmer requires significant training and credentials, many of which are costly to obtain – a fact that one African American farmer we interviewed saw as a barrier to entry (Interview 29). One farmer described the complicated process of getting hired, as well as the working conditions:

> Most people do a couple of apprenticeships before they are ready to run a farm on their own, because it's a lot to learn. So I came in with, I had worked on my college farm for two summers, then I came to this apprenticeship ... And we're all on the farm between 45 and 70 hours a week. (Interview 15)

Urban agriculture presents different opportunities than a traditional – and potentially more stable – job can provide. An urban grower expressed disappointment with office jobs, and knowledge of what was happening in Philadelphia:

> I really want to get my hands dirty. Urban agriculture wasn't always in mind. I discovered it through reading. I had a pulse on Philly and I saw there is an interesting movement in Philly. I am still on the high of it with endless possibilities. (Interview 25)

Some of the farmers we interviewed saw urban agriculture as the next dot.com boom, or a way for young people to drive an industry and revolutionize everyday thinking about food systems and cities. It is striking that some of the farmers draw parallels between urban agriculture and the dot.com boom, given that the economics of urban growing and the new tech economy are worlds apart. However, our interest is in their perceptions rather than the reality. Many of the young growers we spoke to saw themselves as innovators and social entrepreneurs. In the same way that the dot.com revolution changed forever how we interact with technology, they talked about urban agriculture as revolutionizing the way we look at urban land and feed communities. Several urban farmers had become minor celebrities because of the success of their urban gardens and farms. They were giving lectures at college campuses and making TED talks.

The growers were also cultivating new visions of the city, one in which the built environment would no longer always be just concrete and steel. Greening the urban landscape can be key to developing a more sustainable future. Many of them talked about how past urban and suburban development models were outdated, dependent on fossil fuels, and *not the future*. One farmer described being frustrated with the current development and food patterns and the hope of urban farming in the city:

> I saw the greenhouses in an urban setting, and teaching people how to grow is a lot of fun for me. Seeing vacant land being unused, while I watched the suburbs where I grew up being destroyed ... It's the best farmland in the world and we're putting houses on that. So that drives me crazy. It's kind of like a way of getting back at that, is to take the vacant land in the city. If they're gonna be developing unwisely in the suburbs, let's try to do something better with the land that we have here. (Interview 12)

For many of our interviewees, growing food in the city was a way to make a statement about their frustration with our current development patterns, the industrialized food system, and the career opportunities open to them. Most urban farmers and gardeners are well versed in the language of food miles and ecological footprints. For them, urban agriculture is a solution. They feel they are helping make Philadelphia a greener, healthier, more sustainable, and more equitable city. Some of the impacts that interviewees highlight are: improving neighbourhoods by cleaning up vacant, abandoned properties; using unused land in productive ways; growing food for people; and teaching kids about growing, nutrition, and business. While these are all lofty goals, the long-term effects of these activities still need to be tested and questioned. Also, it needs to be asked whether these urban farms are actually promoting equity, particularly at a time when some city officials still view urban agriculture as a means to "stabilize" neighbourhoods to prepare them for development. Also underexplored is the impact of the encroachment of young, mostly white people into primarily African American communities (see below).

A Steep Learning Curve:
The Challenges of Making a Career

While there is excitement about urban agriculture in Philadelphia, there is also recognition that growing food in the city presents a host of challenges and complications. For most urban growers, the learning curve is steep. Many of the young people we interviewed did not grow up farming or gardening. One urban farmer noted the irony that many people who had grown up on farms were not interested in urban farming as a career:

> Many times living in a city, it's always going to attract people who aren't necessarily farmers or people that came from big farming background ... We're looking at it as this exciting, fun, cool thing to do. Whereas you're growing up, and all you have to do is weed every afternoon, like this stinks, this is not fun, and you kinda want to get away from it. Where it's funny, because since I didn't grow up with it, it was very exciting for me because of the kind of work. (Interview 11)

That growing food is "exciting, fun," and a "cool thing to do" was a common theme among urban growers. Growers described the new interest in urban farming as a legitimate career:

> I think it's really weird right now. I feel like I became interested in farming right before the farming frenzy erupted. And it's been really weird to see it become something that's super cool. When I started in college, everyone was kind of like "Oh, what an interesting thing for a college student to do for a summer, oh this must be very relaxing" … When I said I wanted to be a farmer, my parents were like "What, you can't be a farmer, you're supposed to get a white collar job." It was like a blue collar thing when I was doing this upward mobility thing. And I feel like … It's kind of changed a little bit so I feel like my mom is much more proud of me now than she was at the time. At the time she was like "Uh, you are too smart to be a farmer." (Interview 15)

The idea that farming the city is a career goal for some educated young people from mostly middle-class families speaks to the changing perception of urban farming and its growing professionalization. It may also speak to the growing acknowledgment in the mainstream media (e.g., in Michael Pollan and Mark Bittman's *New York Times* columns) that our industrial food system is fundamentally flawed. But whether urban agriculture can actually be a profession was hotly debated among our interviewees. There are no doubt successful farms in Philadelphia, but most of the small-scale growers we interviewed recognized that they were not making enough money from growing food on their lots. Even so, they were excited about their work (much of it done on a volunteer basis), and they hoped to eventually make it more financially sustainable. One urban farmer who started a farm that employed young people recognized that despite its financial limitations, the experience of starting a farm was in itself valuable:

> You must think I'm crazy, but you couldn't ask for a better internship. I'm running a program. I can dream anything I want there, which a lot of people don't have that ability. I could make all kinds of mistakes.

> I have all the glory and not much capital building. I don't get paid for it.
> We learned a lot. (Interview 11)

This farmer was excited about the opportunities to experiment that urban agriculture offers; but also recognized that an inherent challenge with urban agriculture is its longer-term economic sustainability:

> The biggest thing that stands in my way is figuring how I can get paid money to do this. People have had ideas of trying to sell as much food in a day and get a cut of everything's that happening. I wouldn't make enough off of that. (Interview 11)

This farmer highlighted the disconnect between the romantic, popularized view of urban agriculture and the reality. Farming in the city can be gruelling work and may not lead to long-term employment, but in the case of this farm, it provided important job training for unemployed and underemployed youth and helped them secure other positions:

> We know that they're not gonna be working there till they are thirty or forty. So we actually have had a lot of success where we've seen young people leave our program last year and go get jobs somewhere else, and the reason they got better paying jobs and were hired was because they got better through our program. It has this ripple effect that works down the line. (Interview 11)

Several urban farming organizations that emphasized youth involvement focused on the transferrable skills that farm internships could provide, rather than emphasizing urban farming as a career path.

As of 2017, economic conditions have changed so that there are more job opportunities for young people graduating from college. Some of the enthusiasm for urban agriculture as a profession already seems to have waned, now that it has become clear that making a living from urban agriculture can be a challenge. A number of our interviewees no longer manage gardens and farms. They have transitioned to non-profits instead of trying to directly earn a living from urban agriculture. Perhaps growing in the city is more difficult than originally thought and the dream of making ends meet doing so is something for a lucky few.

Civic and Social Justice in the Context of Race, Class, and Gentrification

Notwithstanding the narrative many new urban growers recite about the civic and social justice aspects of their involvement in urban agriculture, there are a number of underacknowledged racial and socio-economic tensions. Discussing race and class is always difficult, and we have tried to be mindful of this as we write about these issues. This is in itself is an enormous research area that deserves a book of its own; we will try here, though, to raise some key questions about the way that urban agriculture in Philadelphia connects to issues of race and class. In 2011 there was little public discussion of race, class, and gentrification in the interviews we conducted with the younger, primarily white growers, even when we tried to press the issue. Clearly, these topics made many growers uncomfortable, and most growers were only beginning to recognize and confront these issues in their work. When we conducted follow-up interviews in 2016, race and class and neighbourhood self-determination were at the forefront of our conversations; this marked an important shift in the way people thought about urban agriculture. The conversation about urban agriculture in the city had evolved, providing us with some insights into how the new urban growers were now connecting with the disadvantaged, disenfranchised, often African American communities where they are working.

One white farmer discussed the changing demographics of people growing in the city and the tension that arises:

> In my opinion, the real problem is that those are completely different demographics. The gardeners that are getting aged out are a lot of

black farmers from the south who have been doing it for a while or first generation immigrants whose kids were not interested. In both those groups, the kids see it as backwards, or going back to farming. The new young people, who are interested and have made farming in the city sexy, are young, white, college educated people. And that's kind of the weird thing to me, behind the current popular moment, as soon as young white people get involved it becomes really cool. Oh look, this new thing happening in the city. It's not new. I think there hasn't, there isn't nearly enough collaboration between those two groups. (Interview 15)

Another white urban grower echoed the generational divide between older African American farmers, younger African Americans who might view urban agriculture as a "white thing," and younger white farmers who were often unaware of the "social and power dynamics." This grower describes this tension where people living in low-income neighborhoods

don't see urban agriculture as attractive to them. Especially the young people in the community, they see it as a "white people" thing. The social and power dynamics of having white people come in and tell them to grow food is problematic because of the history of power relationships. There are a lot of old folks who come from the South and love to grow food in their communities. These same people who are growing food in their communities are also sensitive when white people bring this to their community or tell them that they should be growing food in their communities. There is a long history of slavery overtones. (Interview 32)

The racial and demographic lines around urban agriculture in Philadelphia are complicated and evolving, and there are different visions of the role of gardening and farming. The older farmers and gardeners, many of whom are African American and immigrants, were often a part of an established community farm or garden and were busy maintaining their own plots. They grew food for themselves and gave it to their friends, family, and neighbours and took pride in being able to share their food. Some of the older growers grew up in other parts of the country where farming was part of their tradition. They were not expecting to make a living farming or gardening, and they did not seem to place the same "world changing" expectations on urban agriculture (discussed above). Many of these older gardeners and farmers were connected to established community gardens – some of which had been protected by the NGT – where people had their own plots of land that they tended. When we went to these community gardens they showed us their plots and their vegetables. Unlike the young "new wave of growers", the older growers we met did not discuss plans to create new farms or the potential of turning their gardening hobby into a career. They were certainly interested in discussing gardening resources and strategies for improving their harvest, but many of these growers were not seeking publicity about their garden sites. They were proud of their gardens, happy to be part of a close community, and glad they could share their produce with others, but they did not talk about urban agriculture in the same way that the new generation of growers did.[17]

Navigating Gardens and Farms as Community Spaces

Whether spoken or unspoken, all community gardens and urban farms have relationships with their community. Here it is difficult to generalize about community relationships because each garden and farm is different. In some community gardens, gardeners pay for a plot and garden individually. These gardens are often fenced off or have hours when they are open to the outside community. Others are run more like community farms where people are invited to grow communally. Some gardens are more grassroots efforts, often led by individuals, that do not have formal rules and that encourage people to "hang out." Others are run by non-profits and have a formal educational mission but may also sell food at local farmers' markets. It is common to find signs of "community relations" at the community gardens. Sometimes flowers are planted on the perimeter and signs are posted that explain the benefit of the farm or garden to the community. For instance, at the Emerald

Figure 5. The Emerald Street Community Farm in Kensington invites people to participate.

Street Community Farm in Kensington, a hand-painted sign in Spanish and English welcomes the neighbourhood: "Love is in the veggies! We are a group of neighbors who grow food in our community. Come grow with us." This sign serves as an invitation for the community to participate in the farm.

Plotland, a guerrilla urban farm in gentrifying University City in West Philadelphia, had a pretty painted sign with hearts on it stating: "This is a community garden. Please ask the gardeners before taking fruits and veggies." The point of the sign was to inform neighbours that the plots were being cared for in the community garden. However, the message was not as well received as hoped.

Figure 6. Plotland in West Philadelphia reminds people that it is a community garden.

After this picture was taken, the sign was broken and taken down. In 2017, a new sign has been posted that says "All Are Welcome" and that invites neighbours to join the community garden.

Since urban agriculture can involve anything from large and small community gardens, to private urban farms, to a cleared vacant lot next to a house, to a non-profit community farm designed for supplying local food and educating local children, to an urban farm on school property, and even warehouses where food can be grown indoors (although this is not the focus of this book), urban growers have multiple strategies of community engagement. Sometimes urban growers developed formal programs to engage with residents. Exposing kids to growing food was often a top priority. In other cases, neighbours were invited to garden, or children came over to see what was happening and ended up "hanging out":

> A lot of other people just show up and ask what we'd like them to do whether it's just weeding for eight hours. People just come and work all the time and we just have a great time. Nobody's in charge really, we seem to get a lot of good feedback about when people come to help and volunteer. If they want to just sit around and talk for a few hours, nobody even looks at them weird. We'll sit down and talk with them 'cause that's what we're here for. It's work, but it's also a lot of fun and a lot of knowledge and storytelling. That's kind of how people are involved. (Interview 8)

These reclaimed vacant lots provide places for new growers and people in the neighbourhood to come together. In neighbourhoods that lack green space, where safety is in issue, the garden or farm can serve as a safe space for kids to spend time. Many urban growers talked about kids showing up and becoming very comfortable at the garden or farm. Another urban grower described his relationship with the kids in his community:

> We get tons of kids that just like kind of come back and hang out, swinging on the swing, torture each other in the hammock. There's a father and his son two doors down that have been a huge help from the very beginning. They help move soil, they come. Had a lot of kids help with farmers' market to some degree. It was more so last year. I'm not sure why the dynamic is what it is. But they'll be a group of kids that'll come by every other day for months, then all of the sudden they're not really around anymore. Then they'll be a new group of kids, some are more problematic than others. We have a really good relationship with the neighbors on this block and the upper part of this block here. Some people have been super involved in the whole labor part of things. (Interview 9)

Urban growers are working to develop relationships with the community. As described above, interest in a garden or farm might be high sometimes and low at others, and the farmers and growers are trying to figure out their relationships with the community. For some urban growers, "hanging out" and being involved is a mark of successful community relations; for others, the hope is to inspire people to take action by teaching them to grow and eat healthy food:

> My main focus here is to show people they can grow food [in a vacant lot]. So you bring in soil, use lumber, build it up, and show people what food looks like. Sell it to them for cheaper, cheaper than what they're getting in their neighborhood, even though it's organic and local. We're selling it basically at a loss. The idea is that they'll see this and be inspired to do community gardens in their neighborhood. (Interview 12)

Providing food access, promoting environmental justice through cleaning up vacant lots, and inspiring people to take action in their own communities is important for many urban growers. They see urban agriculture as a tool for modelling how to think differently about their land and their food. Another grower echoes this sentiment:

> That's what urban agriculture is in my mind. You're not gonna get them to change their habits and come and support this farm every day, but you can have them walk by it, and be like, "Oh yeah, Tomatoes are in season." ... It's just these visual things ... They bring their kids for a work day or whatever else ... But I don't see it as part of the economics of the food system at all. (Interview 3)

Many of the new growers saw their work as creating spaces for growing and farming in neighbourhoods. They discussed how they were able to get their neighbours involved in helping to clear the land and how many neighbours are helping to grow food or are actively involved in the project. However, white urban growers tend to downplay the issue of race and class and their role in neighbourhood changes, specifically gentrification. When it came to discussing urban agriculture as a part of "green gentrification," a process where cleaning and greening may lead to gentrification, many younger white growers would point to all the vacant land still available.[18] The city has a huge number of vacant lots, and many urban growers did not feel that Philadelphia had reached the gentrification tipping point (although by 2016, there was more recognition of development pressure). Instead, they saw a void. They were disappointed that the city was not doing enough to improve neighbourhoods, and they felt it was necessary to start their own initiatives. The magnitude of abandonment in Philadelphia is startling to many urban growers, particularly those who are new to Philadelphia, and was one reason why they felt that they had to act. They viewed turning a vacant lot into a garden or farm as a tangible way of making social change. In some ways, the extent of vacancy was attractive in Philadelphia because gardeners and farmers could readily see the need for their contributions in addressing blight and abandonment. A number of the "new wave of growers" we interviewed were motivated to begin reclaiming and cultivating vacant lots in their neighbourhoods after they had lived in a particular neighbourhood for a few years. They became frustrated looking at abandoned lots that were sitting vacant or serving as dumping grounds for old cars and debris.

> I started looking around, talking to people about how I can start my own [garden], because there was all sorts of news coming out about how the city has over 40,000 vacant lots so why not start your own garden on a vacant lot? And I found this particular space ... It turns out almost the entire block is vacant. There's 25 lots that are vacant on this block out of 32 or something like that. (Interview 8, speaker 2)

This interviewee had a plan to develop a garden/farm and was looking for good sites for it.

The disinvested condition of Philadelphia's neighbourhoods provides urban growers with an opportunity to test their grit and to make a difference. The land is available (and up until recently there had been little demand for it), they just need the determination and willpower to clear the land and start a garden or farm. These young urban growers want to experience vacant lot cultivation first-hand and make a difference in a city they perceive as needing help. They make the link between their day-to-day involvement with urban agriculture and larger societal issues. They are proud of their ability to transform the landscape. When asked why they became involved in urban agriculture, they describe themselves as agents of change:

> I am super proud of what we are doing. We'll have hipster friends, people from [the] neighborhood, homeless people. It is fun. You get adapted to neighborhood. You get to see what is wrong with Philadelphia. When you really drive around the whole city you realize this place is a hellhole. (Interview 11)

For young people, many of whom have never farmed before, the prospect of building a farm or garden with their own hands is critically important to them. If the Vacant Lots Cultivation Association initiative from the early 1900s (chapter 2) was about blending ethics with charity, the current interest in urban agriculture revolves around ethics and action. When asked about the motivation to start a garden, one interviewee said:

> I think necessity, interest and just indescribable pride, and just how great it was to get going in this thing ... We're pretty proactive people. So we were just like, well let's start a garden. (Interview 18)

The notion of being "proactive people" was common among the "new wave of growers" we interviewed. Here it is worth questioning some of the assumptions they were making when they framed their role in this way. A white farmer activist recognized the troubling dynamic at play when it was perceived that white people were somehow "saving" the community:

> Going into someone else's community can be really damaging. The result is that we reinforce this perspective among young African Americans: the idea that urban agriculture is a "white people's thing and not for us." The narrative that "communities of colour need white people to come in and save them." This leads to a negative relationship. It reinforces it. Taking a vacant lot and growing food can be good, but

if the neighbours are not included these situations do more damage socially politically than they cause good. We see this all over Philadelphia
and all over cities. (Interview 32)

This recognition that urban farming has the potential to replicate
socio-economic and racial inequalities emerged over the course of
the research project, indicating some promising critical reflection
on urban agriculture. Another young, white urban farmer recognized that good intentions do not always lead to good outcomes:

I had so many people, especially college students, white students, being so pumped about gardens, and I fit that description too. It is great
that they are excited, but there are real implications as to the presence
in the community. There are important power dynamics to consider
when they are going forward and doing this work. You can have good
intentions and a problematic impact. Those two things can exist at the
same time. And it is challenging. Many years into my prior job [at an
urban farm], a kid who I had built really good relations with and his
family said to me "he thought only white people could garden." If that
is the impact that I am having with my work, I am ready to get out of
there. That is just reaffirming a lot of problematic power structures in
our society. That is not the goal! I am still grappling with what niche
I can fit in that is useful in terms of building a more liberated society.
(Interview 36)

This white farmer's experiences and introspection indicate the potential for a more open discussion about what role young white
farmers play in primarily African American and low-income communities. Another young white farming activist described the need
for young white farmers to recognize that they are outsiders and
are coming into communities with a history of being oppressed.
Instead of coming into communities with a vision, this farmer rec-
ommended a different, more humble approach:

If we come from privilege, we have access to assets, but it is up to us
to take a humble approach and to take the time to begin, slowly and
humbly to build relationships. We need to figure out how we can use

our assets. It might be urban agriculture, but it might be something else. It might not be the sexy, Facebook thing that we were expecting it would be. It might be something else. I wish those types of conversations would happen more. We need to take a humble, inclusive position. (Interview 32)

Kirtrina Baxter and other activists in Soil Generation try to educate outsiders before they come into disinvested neighbourhoods of colour hoping to develop gardens and farms. She often encourages people to connect with existing urban agriculture projects in communities instead of starting their own efforts. She describes how she intervenes to help educate and redirect young, new growers who identify sites in disadvantaged neighbourhoods:

Sometimes we hear from people before they get lots. We hear about them before it happens. People are so excited because there is all this vacant land. They think this is their "playground" … Hell no, this is *not* your playground. We reach out to those people. Then it is constant education during relationship building. The key thing is that we are not saying that you shouldn't be interested in the environment. We are not saying that you shouldn't be interested in creating food pathways for people. We are saying that you need to be more educated and informed about your impact and try to figure out ways that you can be less impactful and support those folks who are already doing this work. So we really encourage people to support work that is already out there that is led by people of colour because there is a whole lot of it. There is all this work happening in the city. You don't need to start your own thing. You really need to fold into what is happening here. Sometimes people are receptive. Sometimes they are not. (Interview, 2 July 2016)

The role that many young, mainly white farmers – especially those who are outsiders – play in disinvested, primarily African American communities is an extremely sensitive topic, one that is difficult for many white growers to discuss. It is complicated and nuanced, and every grower has a slightly different relationship with the community. There is no one way to connect with communities. Many of the white growers we interviewed saw their work as supporting disadvantaged African American communities, and many of them referred to themselves as "part of the community." Many also indicated that they had discussed the creation of the garden or farm and had received support for converting a vacant lot into a growing space, or at least no opposition to their doing so. They mentioned neighbouring residents who were friendly and who voiced their appreciation that the vacant lots were being cared for in some way, as opposed to becoming overgrown or attracting litter. A white grower interviewed in 2016 suggested that "white hipsters" might feel uncomfortable talking about the racial dynamics of their work because they had come to it as a way to address their "white guilt":

> It is a largely well-educated group of white hipsters who are approaching this from a highly academic lens. White guilt is a lot of it. You are eighteen and you come from a largely racially isolated community. Young white people deal with internalized racism and guilt. We get into urban agriculture as a way of coping with that and a way of feeling like we are doing more good than harm … People need to acknowledge their privilege. But we develop these understandings in an academic setting. We are not a part of the communities that are suffering from the injustices that we study. We are people who read books and watch movies about problems taking place in other people's communities and then we decide what we think these communities need because this land is cheap, and there is a lot of it available, and we do our thing there. Developing a heightened sense of awareness is good, but contributing to these types of social, power, and political dynamics [is] not. (Interview 32)

An African American farmer echoed the need for more reflection about the role that white urban growers play in disadvantaged, African American communities:

> White people need to be called out about their work. What are your motivations? I would hope that through this work you would understand the history. It is not about standing on the ground in 2016. We carry historical baggage … the hangover from slavery…Urban agriculture is

a small part of a larger problem: public schools, food related diseases ...
Our role is tiny because the problems are so great. (Interview 29)

This farmer was optimistic that there would be a more open discussion about what it means to work in communities that are *not
your own* and that have a long history of disadvantage and institutionalized racism. Another white farmer talked about how a group
of white urban growers who were members of Soil Generation
were now meeting specifically to discuss race and class issues and
to learn to "challenge the race and class and power structures in
their own organizations and work more effectively to serve different communities and to figure out their roles. That seems like a
lot of forward progress. It is messy and complicated to talk about"
(Interview 36).

chapter six

Putting Growing in the City in Perspective

The narrative around urban agriculture suggests that it can fix many urban problems simultaneously. We have found, however, that it does not radically reshape the city or fundamentally change the food system. The reality is that urban agriculture responds to and is bounded by the city's existing social, economic, and politics. Instead of lauding urban agriculture as an activity that can solve multiple problems simultaneously, we need to put growing in the city in perspective and to acknowledge its strengths and limitations. Urban agriculture can have important community, social, and environmental benefits, but there is often a mismatch between the ambitions of many urban growers and what they can actually accomplish.

Urban gardens and farms are often very small. Many exist in a temporary space, subject to displacement and redevelopment. Most rely on volunteer labour. While these factors are not enough to deter many growers, they often limit the scope of their efforts. Urban agriculture activists find it particularly frustrating that gaining access to land for growing food in the city is such a complicated process (see chapter 4). This restricts their ability to scale urban agriculture up to a level where it could do more to address the multitude of problems facing Philadelphia's residents, particularly low-income residents. At the September 2016 City Council hearing on urban agriculture, Chris Bolden-Newsome, co-director of the Community Farm and Food Resource Center, described how in Southwest Philadelphia, a low-income, primarily African

American community with limited access to fresh food, the difficulty of gaining legal access and security for gardens made it difficult for citizens to grow food and improve their communities:

> It feels like a particular kind of torture to know that you have lots lying empty next to you that there's somehow – that if you attempt to grow on them, you will have no protection for that, and that that could be developed, that could be taken from you, you know. When people just want to do for self, that's the emotion that in our community is what's going to help us survive. So I want to help first with that. Remove the barriers to land access. And then create accessible legal pathways to ownership for folks who are growing on city lots. Because the truth is, some folks have already started growing. They didn't wait for the shot or for the permission or whatever. They started growing.[1]

As discussed in chapter 4, the process of formally accessing and protecting land for growing is usually slow, and it continues to be uncertain. That leaves many urban growers operating under the radar in makeshift gardens without legal protections. In addition, communities that have suffered from disinvestment and vacancy, and that could benefit the most from urban agriculture through access to fresh produce, may be the very communities *without* the political and economic resources necessary to benefit from policy reforms such as the Land Bank.

In Philadelphia there is stiff competition for limited funding to support urban agriculture and most farms and gardens are not commercially viable on their own, nor are they protected from development. As a result, the promise of urban agriculture as a solution to many urban problems has not yet been met. Urban agriculture's success is also being thwarted in Philadelphia by growing pressure for development and by some city officials' unwillingness to "value" the community, economic, and environmental benefits of urban agriculture on par with those of redevelopment. Increasingly, as the economy has picked up, urban agriculture has entered into a complicated relationship with gentrification and the displacement of disadvantaged communities. Given the dynamic in Philadelphia, where urban agriculture is seen as a means for low-income residents to gain access to land in their communities and grow much-needed food, but also as a possible precursor to gentrification, in the course of which younger, often white outsiders move into neighbourhoods, often followed by a wave of developers, it is difficult to make a single statement about the role that urban agriculture plays in communities. Urban growers, particularly young white growers, are struggling to make sense of their roles, which may be complicated, racialized, politicized, and dynamic. However, in a city that is rapidly gentrifying, questioning one's role in urban agriculture is a task for urban growers from all socio-economic and racial backgrounds: What is the role of a young, white urban farmer who aspires to create a farm that would provide fresh produce for neighbourhoods of colour? What is the role of a middle-aged black woman serving as the president of her well-established local community garden? What is the role of a community organization seeking to create a farm to employ local youth for (transferable) job training? We found that regardless of the identity, background, and role of different growers, many came up against the same set of challenges revolving around financial (in)stability, uncertain land tenure, contaminated soil, and limited food production potential. When urban agriculture was unable to fully keep its many promises, urban growers argued for a more nuanced understanding of the benefits of urban agriculture; they also saw a need to develop legal and financial support for growing in the city.

The Promise of Financial Stability, Local Food, Green Jobs, and Community Benefits

A diverse group of gardeners and farmers from all socio-economic and racial groups are passionate about social change and optimistic about transforming vacant land and growing local food. Some urban growers see urban agriculture as an important part of the food system – particularly as a way to supplement the diets of low-income residents – as well as a way to show city dwellers, from all socio-economic and racial groups, how food is grown. Yet for all their optimism, they recognize that small-scale urban agriculture cannot compete with larger farms and is often not, on its own or without foundation, city, or business support, economically sustainable:

> The fact is, you can't grow enough food in this city to feed really anybody. A lot of people have been working, trying to have gardening to be taken out of the category of a hobby and into a more legitimatized profession or at least vocation. The challenge is, you can't make any money out of it. You can call it a job, and whatever else, and pay for it by grants, but at a certain point, it's not something that's gonna pay you ... But unless you're selling to the highest end consumer for the highest price, you can't really make any money. (Interview 3)

The vision of urban agriculture revolutionizing Philadelphia's green economy has yet to become a reality. Urban agriculture entrepreneurs are finding that on their own, without support from non-profits and educational groups and without a distributional business model, urban farms cannot produce many living-wage jobs. One of the few farms in Philadelphia that is universally considered to be economically viable is Greensgrow Farm in the Kensington section of Philadelphia, an established farm that has become profitable by diversifying its services and products and by acting as a distribution hub. In addition to growing on site, Greensgrow sells to restaurants and acts as a regional distributor selling plants and produce grown in other locations.[2] Some urban farmers are increasingly looking to high-end food production (such as mushrooms and, more recently, vertical hydroponics) and even vertical farms in former warehouses; others, though, question how selling high-end food meets social and environmental justice goals. Focusing on high-end markets may be a way to create green jobs and economic sustainability, but this approach to urban agriculture is not necessarily about reducing food deserts, creating community green spaces, or achieving environmental or food justice, goals that are frequently cited as the benefits of urban agriculture. Some urban growers justify growing high-end micro-greens for restaurants as a way to support their more community-focused work through cross-subsidization. Others offer that any activity that creates jobs in disinvested communities in Philadelphia is a positive. They argue that Philadelphia needs jobs, development, and tax revenue and that if urban agriculture can provide any of those, it is valuable and needs to be supported by city policies. In addition, managers of farms and gardens that work with underserved youth and sell to high-end restaurants argue that in selling to restaurants they are teaching kids how to be entrepreneurs. That is, they are transmitting valuable business and life skills and thereby preparing young people for the workforce.

One urban grower argued that the promise that urban agriculture will create jobs and compete with more conventional food production perhaps needs to be thrown away:

> It's not going to produce jobs ever, unless it's something where you start a non-profit like this. People think they can grow food in the city and make money, and when you're competing against commercial agriculture that gets government subsidies for X, Y and Z, and then uses immigrant labor and has thirty acres, you're never gonna compete with that. (Interview 12)

This grower suggested that we embrace more realistic expectations for urban agriculture, meanwhile developing financing mechanisms that would sustain it:

> Especially if you're selling to your neighbors in a low-income neighborhood, you'll never make money off that. You're always going to lose money. Never compete with commercial agriculture for anything. It's just the brutal truth. You will always lose money so you have to subsidize it with donations or grants. That's the only way. It seems like people think that they can do it. It's impossible. It's hard enough to have a farm in the country and make money. [Urban agriculture] is great for feeding people and teaching them where food comes from and for greening neighborhoods, and that's about it. It's not jobs, unless you can start another non-profit where you can link up a bunch of community gardens and get grants to feed either your neighbors or to sell it. (Interview 12)

Growers argued for a more nuanced understanding of what urban agriculture can and can't do. Urban agriculture can provide some fresh food, and it can help educate people about how food and food systems contribute to greening the city. But the notion of "teaching" people about food itself needs to be critiqued, for as mentioned in the previous chapter, when (often) young white people come into disadvantaged African American communities and begin telling them what they should and should not eat and how they should grow and prepare food, this can reinforce existing power imbalances or create new ones.

So if urban agriculture does not generate enough food to feed a population and does not pay enough for a career, what is it about? One urban growing advocate was relieved that the conversation was shifting away from profitability, given that urban gardens and farms had so many other valuable community uses:

> I am very heartened to see the discussion moving away from "how can we make urban farming profitable?" because food is not valued enough to make farming profitable, period. That is why you have industrial agriculture now. It is very hard to undercut that ... Most of the farms are nonprofits and rely on grant money and donations and I don't think that is a problem because they provide so many important public services: space for people to gather that is safe, improved air quality, lower crime, access to healthy food. Space that people can take ownership of. The intergenerational connection. We don't ask other public amenities to be profitable so I think it is pretty crazy that they think that these ones need to be. It should be really a bonus if you are able to make some income. (Interview 36)

As noted above, an often underappreciated benefit of urban agriculture has to do with the role it plays in shaping neighbourhoods and its connection to community empowerment and control. By gaining access to land and participating in urban agriculture, community gardeners and growers learn how to self-determine land use in their communities. Kirtrina Baxter from Soil Generation wants people to measure the impact of urban agriculture through a racial justice lens:

What is missing in the critique about urban agriculture is that it is not just about food, but it is about land too. So what we also talk about in Soil Generation is how important land is. When you control the land you can confront the other issues within the community from a place of power. We come from a racial justice lens: we can't not think about all the other ills that are happening in our communities. We focus on these things here because we understand that legal access to land and the ability to self-determine what happens on land can affect all these other issues that we are having in our communities. So that is why urban ag. is so important – because land is the basis for all of it. If you look at urban agriculture without a land analysis, then you are missing a giant piece of the puzzle. This is why urban agriculture can help communities self-determine. It is the land issue. (Interview, 2 July 2016)

Especially for people in disinvested and disenfranchised communities, urban agriculture can be a powerful tool for community organizing around land. Many growers understood that their work in communities was about much more than food:

Every time I go to these food panels, my whole thing is, this is not a food issue. This is a green space, land access issue. We need places in the city that are flexible green spaces, like when a community arises that wants to grow food on a piece of it, great let them do that. If someone wants to have a playground, great, let them do that. But they need to have community control over feeling like all those options are on the table. But right now, that's not where we're going with any of it. (Interview 3)

When we see urban agriculture as a part of a more integrated approach to greening the city and to community self-determination, we measure its success very differently.

The Sustainability of Urban Agriculture in an Uncertain Future

Since many urban growers find it difficult to make a profit from their farms and gardens, they are coming up with other ways to make sense of their role and think about their future. Often this means making peace with the transient nature of the activity. One urban farmer, who developed a farm on a vacant lot, talked about his time horizon:

> I sort of thought about it like, we'll prove ourselves by doing this and improving this, and then hopefully they'll see what we've done and then maybe we can get control over it? Our whole existence is tenuous... Long term, what will happen here? I have no idea ... I don't know, the whole thing, maybe in two years it all goes back to what it was. (Interview 9)

Given the amount of time, love, and capital this farmer invested in this farm, it is surprising to hear that the grower's plans for it were uncertain. In some ways, this perspective embraced the reality that urban farms are stopgap solutions to the vacancy issue. Without land tenure, the future of these gardens and farms is uncertain, and there is a sense of real loss when people who have made a commitment to their communities lose their gardens and farms. This raises concerns for urban growers and city officials alike. However, city officials, who have seen many farms and gardens fail as people lose interest, are reluctant to give land tenure to growers who are "just trying it out" and who do not have a track record of success (Interview 24). They are much more comfortable with established community gardens where the growers are from the community and have a track record of being involved. As mentioned in chapter 4, there are increasingly nods towards creating more permanent urban farms and gardens (through purchase and longer-term leases) through the NGT, Parks and Recreation, and the Land Bank; however, the process is still slow and uncertain and is highly political. Developing acceptable criteria for preserving land for gardens and farms is an important next step. The NGT has developed a Priority Acquisition Plan (March 2016) that identifies gardens and farms in need of protection.[3] Prioritizing which gardens to protect first is critical. However, coming up with criteria to value gardens may be a challenge, because those criteria will need to recognize the diverse roles that gardens and farms play in different communities. One interviewee expressed concern that in the past, the gardens that were "saved" were those with enough resources and community mobilization to attract the attention of the city councillor – assuming, of course, that the city councillor was sympathetic. Not surprisingly, several of the newly protected gardens were in parts of the city that were already gentrifying when the gardens were threatened. While it is important to protect gardens in gentrifying neighbourhoods, there has been some concern that in many more disinvested parts of the city, the city does not know necessarily know about the good community work being done, with the result that gardens are being lost:

> The vast array of community-led gardens are not on the city's radar. For me, that is, "Are they in communities that get any attention from the city?" You know where the investment is primarily concentrated. Where the economic investment is already occurring, the city piggy backs on top of that. (Interview 36)

Some growers express discomfort with the uncertain land tenure system in Philadelphia and have started to question whether urban growing is worth pursuing. One farmer described his ambivalence about supporting urban agriculture when its future is uncertain:

> It's really hard to tease out what's long-term viable. We're really torn for those of us who have been doing this for a while, wanting to support all this movement, all this energy around local food, while at the same time having this historically balanced perspective, that this isn't really sustainable, so how much do we want to keep inflating this bubble, knowing that it's going to help us in the short term without trying to predict a negative future but also, none of these farms are financially sustainable. Most of them don't have long-term land access, and so what is this going to look like ten years from now when local food is no longer the hot topic and we're on to methane, or probably back to coal or whatever? (Interview 3)

There is an uncomfortable recognition among some growers that unless the city makes a more formal commitment to urban agriculture and there is increased financial and legal support for it, its future in the city will be uncertain. Some community gardens have existed since the 1970s, but for the many new farms and gardens that rely on an enthusiastic volunteer base, an absentee or compliant landlord, and a receptive community, growing on that property may be something that only makes sense during economic downturns when land is readily available. Figuring out how to protect gardens and farms when there are increasing gentrification pressures will be a future challenge.

Some interviewees were open about how difficult urban farming really is: urban agriculture is less glamorous than the rhetoric that surrounds it would suggest. Farming is labour intensive and requires toiling in the sun in the summer. Once new growers begin to experience the realities of farming, they may not want to do the work, especially if it means only short-term results that could be easily erased if the land is redeveloped:

> People have these wonderful feel good ideas about the whole urban agriculture movement and all of this, and "let's start a community garden," and "I'm going to save this neighborhood," and it's like it is a huge time commitment, and it's extremely labor intensive. And it's hot in the summer. And it's not like on the hot days you can go to the mall instead – that's when you have to be out there watering things. (Interview 9)

Given how committed urban growers need to be and how challenging it is to make a living from growing in the city, it is not surprising that some guerrilla gardeners and farmers who were initially enthusiastic about transforming vacant lots into farms and gardens have moved on to other things. By 2016, a number of the younger white farmers we interviewed in 2011 had moved on from urban agriculture and had found jobs with urban greening non-profits or with the city. One reason people gave for why they had "given up" hope about the role urban agriculture could play was frustration with the city's slow pace of supporting urban agriculture. One young farmer admitted: "I've gotten a lot older in the last year. All the idealist stuff about the urban farming thing finally hit the wall about the reality of doing any business in Phila-delphia. The whole land access issue is a problem" (Interview 11). In some cases, their farms and gardens were no longer in opera-tion; others were no longer involved in urban agriculture, or they now saw it as something they did in their spare time rather than something that could sustain them financially. In the space of a few years, many of them had traded in the hopes and dreams of urban agriculture for more stable careers with steady paycheques. Urban growers increasingly recognize that there is a high rate of failure and turnover in urban farming and that without outside support these efforts may have a short lifespan: "For every five people that say they want to garden, maybe one will actually make it. I think they need support" (Interview 12).

While it was clear that growing in the city is challenging, many growers were also thankful to the organizations and institutions that offer support to urban agriculture in the form of grants, materi-als, and technical expertise (see Chapter 4). Historically, many com-munity gardens relied on support from the PHS or Pennsylvania State University's agricultural extension program to maintain the garden through the years. Recently, however, with more and more young people showing interest in starting farms and gardens (and seeing if they can turn these efforts into operations that pay living wages), competition has developed – and sometimes even jealousy –for outside support and grant money. One farmer described the competition and was surprised by

> how cutthroat it is for people to get funding. Some organizations are
> much better at getting attention and money, even if they don't know
> what the hell they're doing with growing stuff. I'm not gonna name
> names, but there's certain people or groups will get tons of money, and
> then they can't grow anything. There are people that have these high
> tunnels that don't have anything in it. They are like food activists, but
> they don't know how to actually grow food. I'm more of a farmer, and
> less of an activist. I mean, I really care strongly about a lot of stuff, but
> when you're growing food successfully, or reasonably successfully, you
> don't have time to do a lot of this stuff. So it's kind of hard to do both.
> (Interview 12)

The competition among growers cuts different ways. "Unfortunately," remarked one urban farmer, "urban farmers have jealousy issues because there is a limited amount of resources that we are all fighting for. Everyone is trying to get the same grants" (Interview 11). In this competitive environment, urban growers apply different methods when measuring the success of urban agriculture efforts. Some of the criteria for success include the quality of the food grown, the quantity of food, the agricultural practices used, the amount of money made, the number of grants won, the benefits brought to the community, and the authenticity of community engagement. The interviews did not uncover any agreed-upon criteria.

A number of organizations in Philadelphia help farmers and gardeners directly connect urban agriculture with the food justice movement. The PHS's City Harvest program links community gardens and urban farms to food pantries throughout the city by setting up boxes for pickup. The Food Trust, another Philadelphia non-profit, whose focus is on providing healthy food, particularly to low-income residents, has developed a Healthy Corner Store initiative where fresh produce is sold at neighbourhood corner stores for people who do not live within walking distance of a grocery store. Working in partnership with the Philadelphia Department of Public Health's *Get Healthy Philly*, the Food Trust has helped six hundred corner stores provide fresh produce options.[4] The Food Trust and Get Healthy Philly also manage twenty-two farmers' markets across the city, which increases access to local and fresh food.[5] To make the food sold at the farmers' markets more accessible, the Philly Food Bucks program, another partnership between the Food Trust and the Philadelphia Department of Public Health, encourages recipients of food stamps (known as SNAP) to shop at local farmers' markets, where, after they spend $5, they receive a $2 bonus that can be spent on fresh produce.[6] Farm to City operates fifteen farmers' markets throughout Philadelphia's neighbourhoods, which sell locally sourced farm products.[7] Programs that bring fresh food to low-income communities that have traditionally lacked grocery stores and access to fresh food are important for changing the food system. These programs help make urban agriculture a critical piece in larger food system reform.

However, even as people enthuse over the amount of food being grown, shared, and sold in low-income communities, there are concerns about the quality of the soil and the food itself. This presents a potentially critical environmental justice issue – how to provide underserved neighbourhoods with fresh *uncontaminated* produce. Many growers we interviewed acknowledged the limited capacity and unknown character of urban soils. They usually opt for raised beds to ensure that their food is not contaminated. The interviewees were unclear, though, regarding what the best policy (besides raised beds) would be to deal with soil contamination. Responses to questions about food safety were surprisingly underdeveloped, given that people were eating the food they were growing and in some cases selling it. One farmer who sells vegetables to local restaurants admitted: "Yeah, we tested the soil. You know, it wasn't good soil but there were no heavy metals" (Interview 16). Other farmers found soil testing too expensive and did not test their soil. Some growers remediated the soil with certain plants and mushrooms that uptake toxic compounds. Others are careful about what types of fruits and vegetables they grow. They know that some plants bio-accumulate more than others. There is also the question of children's exposure to the contaminated soil (particularly lead exposure) between the raised beds when they play while their caretakers are working. As discussed in chapter 4, in 2014 the Food Policy Advisory Council (FPAC) developed a Soil Safety Working Group to develop policies on how to test for and manage soil safety.[8] Much of the advice involved conducting a site history before growing anything on the site, then testing the soil and afterwards using best growing practices. While it is good news that the city is taking soil safety seriously, this issue requires much more research and action.

A More Nuanced View

To put urban agriculture in Philadelphia in perspective requires us to take a more nuanced view of the benefits it can and should provide. A key question many urban growers wrestled with was the value added of food production in the city, given that Philadelphia is so close to Lancaster County, the breadbasket of the region, and given that the soil in Philadelphia is contaminated. Some urban growers suggested the interest in urban agriculture and fresh healthy food should be used to develop a link between the city and its rural environments, with less emphasis on urban agriculture as the best solution to an incomplete food system. Other growers felt that urban farms and gardens would continue to play a vital role in transforming vacant lots from eyesores into productive spaces. Others, activists involved in Soil Generation, saw a focus on urban agriculture as a way of building community power and self-determination. Still other growers found urban farms to be important as demonstration projects. They show people where food comes from, even if the farms may not be able to provide significant quantities of food. Different stakeholders conceived of urban agriculture differently: it was providing food, helping empower communities, and serving as a force for social change. For cities, urban agriculture can be an important tool to engage communities in reshaping neighbourhoods, but if we want urban agriculture to have more impact, it needs to be valued and supported. In Philadelphia, urban growers saw missed opportunities for urban agriculture, and that is why many of them were so actively involved in changing the policies around urban agriculture at the city level. They felt that if urban agriculture in the city was valued and supported, it would be better able live up to expectations.

chapter seven

Transitioning Towards a Sustainable Future?

This book has reflected on urban agriculture and its role in an older industrial American city attempting to reinvent itself as the "Greenest City in America."[1] Signs of urban agriculture appear just about everywhere in Philadelphia: there are community gardens across the city, there are urban farms, promotional materials are posted at City Hall, local museums hold talks, gardens pop up in vacant lots, university student associations host events, a local sustainability magazine prints articles on urban growing, farmers' markets are multiplying, and informal gardens in vacant lots dot the urban landscape. The ubiquitous presence of this activity raises important questions about the hopes and promises of urban agriculture for a city like Philadelphia, which grapples with vacancy, crime, poverty, food insecurity, and unemployment but at the same time is increasingly gentrifying, and where officials are pushing to develop a sustainable, attractive, and successful city. Urban agriculture fits into the sustainability framework, but is it just a fad? Is it a nice way for a city to improve its aesthetics temporarily while we wait for the next wave of development? Or is it something different? Is it mainly about young, altruistic millennials changing the way we think about urban space and our food systems? Is it new? Can it be scaled up to change the way we think about urban land? Or is it part of a takeover of low-income African American communities by young, white millennials who fail to acknowledge their own privilege? Or, alternatively, can it be a means for people in disinvested communities, particularly people of colour, to build community power? Or is it something in between? There is plenty of enthusiasm for urban agriculture across the city; however, there is considerable disagreement about what exactly the role of urban agriculture should be. Where should the "green wave" that urban agriculture is part of take a city like Philadelphia? And can other cities learn from Philadelphia's experience? This book has explored many of these issues. In this chapter we offer a summary of what we see as the main findings.

Main Findings

Urban Agriculture Is NOT New

Despite the current sense of newness and urgency surrounding growing in the city, farming and gardening have long been mainstays in Philadelphia and many other cities. Some argue that urban agriculture was part of William Penn's vision for a successful Philadelphia in the 1600s. While Penn's instructions for an agricultural city did not transpire as he envisioned, farms and gardens have cropped up throughout the years, particularly during economic crises. Some gardens provided food during economic depressions and wars, while other provided space for community gathering. Some gardens were officially approved through land leases or bought through a land trust, while others were guerrilla gardens maintained by committed gardeners. Some of these gardens have lasted a year, while others have lasted multiple decades. An untold number of gardens have been lost to development. They came and went without much fanfare.

A Growing Interest in Formalizing Urban Agriculture and Connection with Sustainability Is New

From their inception in the 1890s, urban farming programs in Philadelphia were conceived as mechanisms for low-cost vacant land management. This left farms with temporary and informal status in city land policy and vulnerable to displacement from

development pressures. Today, similar to urban farming advocates of years past, urban growers in Philadelphia are contributing
countless hours of labour and energy over multiple years to maintain garden and farms sites across the city with limited support or
resources. Yet something new is happening in Philadelphia around
urban agriculture. Whereas in the past, urban agriculture was
seen as an informal use or was supported by non-profits like the
PHS and NGA, today urban agriculture is being institutionalized
through city sustainability and land policies in a way it was not
before. The institutionalization of urban farming in Philadelphia
suggests that it may become a more permanent feature in the urban
environment, as opposed to a stopgap measure during periods of
economic decline and limited resources, as in the past. This trend
towards acknowledging the multiple benefits of urban agriculture
and supporting the institutionalization of urban agriculture in city
planning documents and land disposition policies is directly linked
to the sustainability movement in Philadelphia, the city's adoption
of its first formal sustainability plan, and the motivation and efforts of a network of urban agriculture activists. Urban agriculture
is also serving as a "draw" to the city for millennials and the "creative class," and this interest in urban agriculture is seen by city
officials as a key part of becoming the "Greenest City in America."[2]
Over the past ten years, urban farming has been recognized as a
community amenity and as an activity that supports a sustainable
urban lifestyle. We see this in the way that people talk about urban
agriculture as being "cool" and as solving numerous environmental and social challenges. Growing food in the city is also viewed
by advocates as a "fix" both for the food system and for disinvested
neighbourhoods. Farmers reduce their food miles, enjoy fresh and
hyper-local produce, and contribute to urban redevelopment in
neighbourhoods challenged in part by high rates of vacancy. The
view that growing in cities is integral to sustainable urban living is,
in part, driving the new interest in urban agriculture and the willingness of city officials to formalize it.

   However, what gives urban farms and gardens their opportunity for permanence is the city's recognition that urban agriculture
can help "reposition and repurpose Philadelphia as a city of the

future."[3] The Mayor's Office of Sustainability's identification of urban farms as a way to meet one of fifteen sustainability targets (Target 10: "bring local food within a 10 minute walk of 75 percent of residents") provided the baseline support for lifting urban farming from a stopgap measure to an important part of the path towards a future, more sustainable Philadelphia.[4] This support has led to a series of policy initiatives that have embedded urban agriculture in city policies; those initiatives include a City Food Charter, the Philadelphia Food Policy Advisory Council (FPAC), a comprehensive plan and zoning code that acknowledges urban agriculture, a new Land Bank (which is partnering with the NGT to identify garden sites for protection), and plans for a Philadelphia Urban Agriculture Strategic Plan. There have also been numerous efforts by city agencies – some successful and some not – to support urban agriculture through specific projects and policy changes. However, here it is important to note that many of the policy inroads have been achieved only because of the strong advocacy work being done by a network of activists who have been making the case that important community, environmental, and economic benefits come from developing and preserving urban agriculture in the city. At every turn, these activists have been there to lobby for urban agriculture to be included in city policy-making.

Despite this progress towards integration, the institutionalization of urban agriculture poses challenges for all stakeholders involved and is very much an ongoing process. Not everyone views it as a panacea. Urban agriculture competes for land that can be used for redevelopment, and in a city with high rates of vacancy and poverty, redevelopment and the return of properties to the tax rolls is an important goal. In addition, there is growing concern about what happens when young, white urban growers take over vacant land in low-income African American communities, especially now that many of these neighbourhoods are already feeling the squeeze of gentrification. We see the ambivalence about urban agriculture in the uncertainty around the city's policies towards it. Many growers interested in gaining access to vacant land have submitted EOIs and later expressed frustration with what they perceive as a bureaucratic and difficult-to-navigate system. In some cases the formal procedures have simply added additional work and barriers; in other cases it seems as though the city lacks the resources or perhaps the will to follow through on its support for urban gardens and farms. In addition, the Land Bank policies, which urban growing activists were actively involved in developing, so far have been slow to protect urban gardens and farms. And there is concern among urban growers that the Land Bank may have come too late and may even help streamline the dislocation of gardens and farms, because of the rapid redevelopment and gentrification the city is now undergoing. That said, there have been some promising developments, including Parks and Recreation's Farm Philly program and the NGT's contract with the Land Bank to identify gardens for preservation.

In many ways the jury is still out on urban agriculture policy in Philadelphia. There has been important "official" progress – for example, urban agriculture is now recognized in city policy documents, and activists have pushed for its inclusion – but at the same time, urban growers are concerned that the city is not completely on board and is, in some cases, failing to effectively implement the policies it has adopted. Also, city councillors wield a lot of power through their councilmanic prerogative, which allows them to veto the use of city land for urban farms and gardens, so they must be brought on board if urban agriculture is to become more than a fringe movement. There was progress on this front with City Council's resolution to hold hearings on urban agriculture. The September 2016 City Council hearing brought a diverse group of more than twenty urban growers to testify about the importance of gardening and farming as a way to provide resources to underresourced communities.[5] City councillors were excited to learn about many of the initiatives across the city. They were particularly interested in the possibility that urban agriculture might be connected to jobs and economic development – an important challenge in Philadelphia, a city that struggles with high unemployment.

The Question of Urban Agriculture Is Unsettled and Contested

The evolving discourse about the role of urban agriculture in Philadelphia makes it clear that there are many different visions of gardening and farming in the city, with growers advocating for long-term land tenure and city officials carefully positioning urban agriculture among other urban sustainability and redevelopment initiatives. Institutionalization remains a frustrating process for activists, yet it is clear that urban agriculture has worked its way into multiple components of city planning and land policy, which suggests that it will be a permanent part of Philadelphia's future.

The institutionalization of urban agriculture may well ensure that urban growing continues in the city; even so, many urban growers recognize that questions of procedure and access must be resolved. The new mechanisms and institutions that have formalized access to vacant land, such as Philly Landworks and the Land Bank, will very likely have an impact on urban agriculture. However, since the stated goal is to streamline access to vacant land, these mechanisms tend to privilege those who have more resources – such as access to the Internet, fiscal resources, and lawyers. As long as there is still a digital divide, and as long as some neighbourhoods are severely disinvested and disenfranchised, access to land cannot be completely fair, for that access relies on the ability to find land "online" in the city's database and then to overcome the many financial and legal barriers to access. Streamlined land acquisition processes in Philadelphia will likely make it much easier for developers to gain access to vacant land. We anticipate that developers, who have far deeper resources, will continue to have an edge over urban agriculture unless urban growers and activists can make a stronger case about the importance of urban agriculture as a means to provide healthy food, educate and empower people, and promote economic development. They need to make the case that land for urban agriculture needs to be prioritized and protected. Currently, while the Land Bank acknowledges "use value" and community benefits in its documentation, there are still concerns that the city's primary goal is still increasing the "exchange" value of land. We also need to examine how the Land Bank's approach to transferring land (and associated wealth) will impact low-income African American communities, particularly in the face of rapid gentrification.

Social Justice and "Browning the Urban Ag." Agenda Need to Be Priorities

From a social justice perspective, urban agriculture has the potential to benefit communities in multiple ways. Growing food in the city is empowering, educational, inspiring, and exciting. Spaces that were previously underused are now producing food. However, urban agriculture activists must pay heed to the sociopolitical, economic, and racial dynamics of their work, because ultimately that work is about "place-making." This can be hard when many proponents of urban agriculture are not from the neighbourhoods where they are working. Activists who see urban agriculture as a solution need to take a step back; this means acknowledging and supporting the existing assets in communities instead of parachuting in and offering solutions. It is important for leadership to come from people in the community and people of colour. Younger, white farmers who have a passion for urban agriculture and want to work in disenfranchised communities need to be more reflective about their roles, more humble, and recognize what it means to come from a place of privilege. They should also recognize that urban agriculture is not a suitable activity or solution for all vacant lots or communities.

In Philadelphia, what Kirtrina Baxter refers to as the "browning the urban ag." agenda needs to happen if urban agriculture is really going to help improve the living conditions and life chances of low-income people of colour. A positive development is that FPAC's membership is becoming more reflective of the city's population, with more representatives of color; however, decision-making around urban agriculture in Philadelphia has yet to adequately reflect the city's demographics. Groups like Soil Generation are playing an important role in "browning urban ag." by building relationships among the "new generation of growers" and existing residents and also by educating policy-makers and activists. Most importantly, they are also empowering low-income African Americans to determine for themselves the use of community land and to navigate the land acquisition process. Community gardens and farms are seen as a mechanism for developing community power.

There are also concerns in Philadelphia that because of councilmanic prerogative, a practice under which city councillors effectively make final land use decisions in their districts, even as activists try to influence policy-making through the FPAC, the Land Bank, and community mobilization, urban agriculture will not succeed until councillors buy into it. To get more political buy-in, urban growers need to demonstrate to policy-makers and city councillors that urban agriculture is a vital and permanent activity in the city, one that provides tangible benefits for disinvested communities by offering them educational opportunities, community empowerment, control over neighbourhood land, access to fresh food and open space, economic development opportunities, and so on.

Growing a Sustainable City Is Challenging and Political

While other studies have examined the direct impacts that urban agriculture has on the environment, such as reducing stormwater run-off and providing fresh produce, this book has focused on how the urban agriculture movement connects with sustainability policy-making and planning. Retrofitting existing city policies to meet sustainability needs has not been an easy task for Philadelphia, nor has it been for other cities. It is very much a work in process. Becoming a more sustainable city requires re-examining existing planning practices and land use policies. In Philadelphia, we see efforts both to *centralize* functions (by creating the Land Bank, the Office of Sustainability, and plans for a Philadelphia Urban Agriculture Strategic Plan) and to *distribute* functions across city agencies, businesses, and the third sector (such as the collaboration between the Land Bank and NGT). It is a challenge to centralize some responsibilities while distributing others, as Philadelphia is doing. Meanwhile, urban agriculture activists are pushing for creative ways to integrate sustainability issues into existing city policies. Incorporating urban agriculture into the zoning code is one way to modify an established planning protocol so that it embraces sustainability. When urban agriculture is brought into the zoning code, it becomes an integral, accepted part of urban land use rather than a fringe use that can only be approved through special permit. However, as we saw with the efforts by a city councillor to change the new zoning code to make urban agriculture more difficult, it is clear that urban agriculture as an acceptable use of land is still contested, despite having been included in formal documents. Activists for urban agriculture have to keep making the case that urban agriculture is vital to a sustainable city, and not just temporary.

City officials are adamant that Philadelphia is *not* Detroit and that the future of the city will involve redevelopment; even so, urban agriculture activists have succeeded at including urban agriculture in citywide policy-making. The city has recognized the importance of promoting and protecting urban agriculture in the city, instead of earmarking all vacant lots for commercial or residential development. We see this in the incorporation of urban agriculture as a use for community benefit in the Land Bank's disposition policies. Of course, city officials do not envision converting *all* vacant lots to farmland or gardens; the point is that they are increasingly open to seeing urban agriculture as an another type of redevelopment. As evidence of this support, the city is partnering with the NGT to identify gardens to protect. For these gardens, the city will provide up to five-year leases to community gardens; this could lower the barrier for land acquisition and tenure. But there is still uncertainty about the process of obtaining legal land access, and the EOI route continues to be slow and uncharted. Overall, urban agriculture policy in Philadelphia is still in flux, and the city still has a long way to go to protect urban gardens and farms from development.

From an environmental perspective, one other surprising way in which urban agriculture is making Philadelphia more sustainable is by uncovering new and potentially controversial issues, such as soil contamination. To address mounting concerns among gardeners and farmers as well as consumers of Philadelphia-grown produce, efforts have been made to convene scientists, government officials, and farmers to discuss establishing citywide soil contamination levels and investing resources in remediation. While most cities recognize that contaminated soil is a concern, that concern is usually reserved for highly contaminated sites, which are designated as brownfields or Superfund sites. Most of the soil in post-industrial cities (including Philadelphia) quite possibly contains some contaminants, such as lead (primarily because of the use of leaded gasoline and lead-based paint), that may pose health and environmental hazards. Urban farming is raising the profile of this ubiquitous yet underreported environmental problem, giving it a place at the sustainability policy table.

In addition, we see an opportunity to better connect the agenda of the Philadelphia Water Department with the urban agriculture agenda. City Council has exempted community gardens from the stormwater fee, and this is a step in the right direction; however, there could be more coordination among Farm Philly, the NGT, the PWD, and other city agencies to develop strategies for using urban agriculture as green infrastructure. This would require more synchronized planning around community amenities and urban greening in the city. Given that the city has an estimated 40,000 vacant lots, there is a lot of opportunity for creative place-making and for connecting green stormwater infrastructure with urban agriculture. There may also be interesting ways to finance urban agriculture to reflect the environmental and economic benefits it provides the city through stormwater management. However, developing this type of integrated funding stream will require significant coordination among city agencies – an ongoing challenge.

Lessons for Other Cities

Philadelphia's experience holds a number of lessons that may be applicable to other post-industrial cities that have vacant lots and active urban farming movements and where there is a strengthening push for sustainability. First, efforts to professionalize urban farming have had limited success in a market sense; most such farms rely on grants to succeed. Urban agriculture has *not* generated numerous well-paid jobs across the city. Many younger farmers envisioned carving out a new, socially responsible career niche in urban farming but quickly ran into a conundrum. Most of them hoped to sell their produce at a low cost to address urban food deserts, but doing so made it even more difficult for their farms to break even. Some farms have experimented with different business models – for example, they have made connections with restaurants that will buy their produce in bulk, or they have branched out into food distribution. Efforts like these expose the reality that farming in the city will remain a challenge because of suboptimal soil, limited land availability, and national food procurement policies that favour cheaper alternatives. This does not mean that urban farming doesn't play an important role in the food system or in the city more generally; the point here is that its true impact, particularly on job creation and food production and access, must not be overpromised.

Second, there has been a lot of policy change in Philadelphia because urban farming and gardening activists have ensured that urban agriculture is included in revised policy documents, from *Greenworks* to the Land Bank. They have mobilized and been extremely vocal. The Garden Justice Legal Initiative (GJLI) has brought legal and mobilizing expertise to the policy conversations around urban agriculture. Policies that institutionalize urban agriculture require grassroots support as well as a constituency; a top-down or top-heavy approach will struggle to succeed. In Philadelphia, the grassroots urban growing activists who are helping shape policies around urban agriculture need to be more representative of African American and other minority communities. Soil Generation has been an important organization when it comes to changing the dynamics and encouraging the "new wave of growers" and city officials to talk about race and class when they are evaluating and promoting urban agriculture. A lesson from Philadelphia is how important it is to contextualize urban agriculture in the city's history and politics: recognizing urban agriculture's long history in the city, the key role that people of colour have played and continue to play, and the role that urban agriculture can and should play in transforming neighbourhoods. For urban agriculture to really become part of a city's meaningful sustainability agenda, it needs to connect to larger conversations about race and equity. Growing in the city needs to offer benefits to low-income residents, many of whom have for years felt forgotten by city officials. Residents of colour from these neighbourhoods who are already growing need to have their efforts and roles recognized. They also need leadership roles in the creation of new urban gardens and farms instead of having these roles filled by outsiders.

Grassroots efforts to promote urban agriculture also need to get buy-in from City Council, particularly because in Philadelphia, the city councillors continue to wield enormous control over land access. We see progress being made in this area with City Council's willingness to hold hearings on urban agriculture and to support the stormwater fee exemption. This is important because until councillors buy into urban agriculture as a valid community use worthy of preservation, one that has widespread community benefits, we will continue to have situations where individual gardens or farms are fighting for land tenure and survival, often begging their councillor for assistance, sometimes to no avail.

Urban agriculture has thrived in a semi-informal state for decades, relying on volunteer work, tenuous tenure arrangements, and often inconsistent resources. This is changing in Philadelphia, where activism around urban agriculture has been instrumental in pushing the city to implement new policies. The increasing enthusiasm for urban agriculture points to the need for a grower-informed approach to developing policies that will support the work of farmers and gardeners from diverse socio-economic and racial backgrounds. For instance, when the city offered one-year leases for community gardens, growers indicated that this tenure length was not appropriate for an activity that relies on the creation of rich soil – a process that can take years. The city responded by allowing for the possibility of multi-year leases (and the possibility of purchasing land) for some community gardens and farms. This extended commitment – though not as widespread or permanent as most farmers and gardeners would have liked, and also a stalled process owing to the lack of response to the EOIs – better represents the type of arrangement required for gardens and farms to be productive. Farmers and gardeners need to make sure that city officials hear their concerns. They also need to frame urban agriculture in such a way that city officials and policy-makers see urban agriculture as worth preserving because it helps them meet other goals. In *Green City, Clean Waters*, the PWD used a "triple bottom line" approach to make a convincing case for the value of green infrastructure in reducing stormwater management costs, raising property values, and providing quantifiable community benefits.[6] However, urban agriculture activists have not systematically quantified the benefits they provide in economic terms, and this leaves them more vulnerable to displacement when the exchange value of land trumps its use value.

Finally, Philadelphia's experience indicates that process equity is important for maintaining an inclusive approach to urban agriculture. Given that urban growers have long been working under the radar, new opportunities for leases, land protection, resources, and recognition are exciting, but careful attention should be paid to *who* benefits from the new system and *who* is involved in decision-making and agenda setting. Also, what are the longer-term community impacts? In Philadelphia, the "new wave of growers"– many of whom are white and well-educated – have helped set much of the agenda around urban agriculture in the city. Many advocates of urban growing have noted this, and there have been important efforts to change the leadership dynamic in the city so as to engage people of colour in decision-making about their communities. However, the disconnect between the "new wave of growers" and the communities where they garden and farm needs much more critical reflection. If the city is committed to urban agriculture as a means of addressing the numerous challenges facing disadvantaged, primarily African American communities, we need policies that will specifically allow for more community control over decisions about community land for gardens and farms. The revival of the NGT is promising, given its long history of working with communities to gain legal right to their community gardens and farms. More diverse coalitions of urban growers and activists, such as Soil Generation, could also help more explicitly connect urban agriculture to issues of social and racial justice and equity.

Of course, growing in the city is an activity that varies from place to place and is supported by different individuals and groups for a wide array of reasons. Also, the role that urban agriculture plays is different in every city and changes over time. For instance, urban agriculture may still be more viable in Philadelphia than in cities with fewer vacant lots and higher property values such as San Francisco and New York. Moreover, the form of urban agriculture itself may change in different cities to meet the demands and challenges those cities face. For instance, in New York and San Francisco, urban agriculture may move from vacant lots to rooftops as those cities become more unaffordable and vacant land becomes scarcer. This may be what happens in Philadelphia twenty years from now once thousands of vacant lots have been redeveloped. Given the different economic, social, physical, and market conditions facing each city, there is no one set of policies or protocols around urban agriculture that can be easily transferred without some adjustment to make them context-specific. Every city must make a concerted effort to develop its own urban agriculture policies so as to address the specific challenges it faces and meet the needs of its diverse communities. An organization like the FPAC is critical. The challenge is to make sure the stakeholders are sufficiently representative of the diverse interest groups and communities that are involved in and affected by urban agriculture. Connecting the recommendations made by groups like the FPAC with the political process is also key. This requires strategic positioning and political acumen. Stakeholders in Philadelphia are realizing that the way to make policy change in Philadelphia is to get the city councillors on board.

But all cities need to ask certain common questions about the role of urban agriculture in urban development. The issue of use versus exchange value of land will need to be debated in inclusive policy forums. The question of whether urban agriculture is a temporary or permanent use will always be contested. Each city's answers to these questions will determine the policies adopted. In all cities where urban agriculture is a goal, planning processes and the resulting plans will need to be updated to promote and sustain urban agriculture. Urban agriculture will need to be supported rather than simply tolerated; it will need to be more than an under-the-radar activity. Government officials will have to engage with growers to meet their needs at the same time that other city goals are being met. There need to be comprehensive planning discussions about how to reuse vacant land as parks, schools, gardens, farms, affordable housing, green infrastructure, commercial space, and even ideas we have not yet explored (solar and wind farms, etc.). Perhaps there are innovative ways to combine these uses – for example, connect urban agriculture with stormwater management and distributed energy production and cross-subsidize all of these. Government officials will also need to check in with community members through formal and informal channels to ensure that their voices are being heard so that urban agriculture actively supports community self-determination. Developing this ongoing conversation about the role of urban agriculture in cities will require engaging urban growers and community members in ongoing visioning processes about the future of the city. Urban growers, who are often used to working off the grid or against or outside the system, will benefit from being engaged in policy discussions about the city's future. Urban growers may also need to be open to the idea that not every vacant lot should be a farm or garden, for city officials have other goals they need to address. City officials will also need to take urban growers seriously and view them as partners rather than as adversaries. For urban growers to be valued as a key part of a sustainable city, there also needs to be more quantifiable research into the community, economic, and environmental benefits of urban agriculture. Urban growers need to be strategic about linking their work to other city goals and working to develop these synergies. Together, city officials and urban growers and their advocates can work together to imagine a future that capitalizes on the benefits associated with urban agriculture. In Philadelphia, many of the pieces are in place for more effective policy development around urban agriculture. The fact that the City Council in 2016 held hearings on urban agriculture is an important development, one that has allowed more voices to be engaged in policy-making and that serves to recognize the important role that urban agriculture can play.

The story of Philadelphia demonstrates that this process of engagement and policy change will take time and will not always be easy. There will be challenging questions regarding how long leases should be for, how to price and allocate land and water resources equitably, and how to ensure that growing practices are safe. There will also be questions about whether new strategies are available to support urban agriculture and make it financially viable. There will also be times when community residents do not want urban gardens and farms, opting for other uses. This may not be what urban growing activists want to hear, but they need to respect community concerns. There will also be questions about whether urban agriculture is truly meeting the needs of local communities and how broadly the benefits of urban agriculture are shared. How urban agriculture fits into the urban sustainability framework also requires asking questions about what cities should look like in the future. As long as cities are to be places where people have local opportunities to eat food they grow, and as long as abandoned spaces are reclaimed as something alive, there will be a place for urban agriculture. However, cities are not blank slates. They have long histories of racial and social struggles. Vacant land in disinvested communities is often a manifestation of these complicated histories. If we are to fully engage in a thoughtful debate about the role of urban agriculture in our cities, there needs to be more recognition that decisions about urban land use are intensely political and that in American cities they often divide along race and class lines. There are winners and losers. In an era when many citizens are worried about gentrification, it is imperative that we have a more open discussion about who the winners and losers are. We also need to better understand the impact that urban agriculture has on neighbourhood change and gentrification. And there will always be competing values and tradeoffs. If part of the city is set aside for urban farming rather than redevelopment, planners and residents need to understand the impacts of that decision on the tax base and on social, environmental, and economic development. To ensure that social and racial justice goals are being met, cities also need to develop formal means for soliciting community input and for engaging residents – particularly from disadvantaged African American communities – in leadership and decision-making roles around agriculture policy and land disposition.

Urban agriculture needs to be included in the broader narrative of sustainability and development, and city officials, growers, and residents need to think critically about what urban agriculture can and can't do for our cities. First, as this book highlights, we need to have realistic hopes for urban agriculture, which cannot be and should not have to be the single fix for all urban problems. We know that currently urban agriculture does not provide anywhere near 100 per cent of a city's food needs and likely will not in the future. Yet there are policy steps that could be taken, such as promoting the local procurement of food, that could help support urban agriculture and help support more just food pathways. It is also unlikely that urban agriculture will be a major job creator. But it will create some jobs, and it can be used for valuable job training. Urban agriculture provides an opportunity to physically transform our cities through hard work, to connect with others, provide food, educate children, and learn about the environment. In the future, urban agriculture may also be a critical part of larger greening and sustainability initiatives aimed at reconnecting the city to nature through connecting with green infrastructure. In the modern world of e-mail, stress, food deserts, processed food, chemicals, and large corporations dominating the food system, urban agriculture provides a small but exciting challenge. It is no wonder that people get so excited by it. Urban growers, through their actions, make a statement about the future they want to see. They don't want to see disinvestment and vacancy. They want kids to grow up understanding that they can plant a seed and watch it grow. They don't want to continue to be disconnected from their food. They want to use the city's resources (vacant land) and their sweat equity to promote change and make our cities better places to live. It is an optimistic and exciting project, but it needs to be more intentional in meeting its stated social justice goals. If we want urban agriculture to be permanent and accepted and to support city goals, we need to recognize the opportunities and challenges it presents. That means not being afraid to ask how urban agriculture will help us grow a sustainable city.

Notes

Chapter One

1 See, Michael A. Nutter, http://www.manutter.com/biography
2 Michael Nutter, *Greenworks 2009*, https://beta.phila.gov/
   media/20160419140515/2009-greenworks-vision.pdf. 7.
3 City of Philadelphia, "Vacant Lot Program," http://www.phila.gov/
   clip/vacantlotprogram/Pages/default.aspx.
4 Alfredo Lubrano, "Phila. rates highest among top 10 cities for deep
   poverty," *Philly.com*, 26 September 2014, http://articles.philly.com/
   2014-09-26/news/54322611_1_deep-poverty-poverty-line-south-
   philadelphia.
5 Pew Charitable Trusts, *Philadelphia: The State of the City: A 2016 Update*,
   http://www.pewtrusts.org/~/media/assets/2016/03/philadelphia_
   the_state_of_the_city_2016.pdf, 1.
6 Ibid., 1.
7 Ibid., 1.
8 Ibid., 1.
9 Philadelphia Land Bank, *Strategic Plan and Performance Report 2017*,
   http://www.philadelphialandbank.org/wordpress/wp-content/
   uploads/2015/01/Land-Bank-Strategic-Plan-to-Council-01-20-17.pdf, 6.
10 Michael Nutter, *Greenworks 2009*, https://beta.phila.gov/media/
   20160419140515/2009-greenworks-vision.pdf.
11 Statistic of 40,000 vacant lots can be found at "City of Philadelphia, Vacant
   Lot Program," http://www.phila.gov/clip/vacantlotprogram/Pages/
   default.aspx.
12 Philadelphia Land Bank, *Strategic Plan and Performance Report 2017*, 10.

13 "Eyes on the street" is an idea from Jane Jacobs's famous book. *The Death and Life of American Cities*. New York: Random House, 1961.

14 Michael Nutter, *Greenworks Philadelphia 2015 Progress Report*, https://alpha.phila.gov/media/20160419140539/2015-greenworks-progress-report.pdf.

15 Max Marin, "After Dodging Sheriff's Sale, Community Gardens Look Ahead to the Land Bank," *Plan Philly*, 25 May 2016, http://planphilly.com/articles/2016/05/25/after-dodging-sheriff-s-sale-community-gardens-look-ahead-to-the-land-bank.

16 We have been questioned about our use of the term of *urban agriculture* because, some argue, the scale of what is happening in the city is not agriculture, it is gardening. It is not our intention here to debate whether it is agriculture or not, but we are using the same terms that we see in the literature, policy reports, and the discourse among activists.

17 The concept of "Farmadelphia" was highlighted in 2005 when Front Studio Architects won the Urban Voids design competition with a plan called "Farmadelphia." For more information about this plan, see http://www.gfcactivatingland.org/explore/ideas/farmadelphia-business-plan/

18 Matthew Doland, "New Detroit Farm Plan Taking Root," *Wall Street Journal*, 6 July 2012, https://www.wsj.com/articles/SB1000142405270230489870457747909039075780.

19 Delaware Valley Regional Planning Commission, 2010, *Greater Philadelphia Food System Study*, http://www.dvrpc.org/reports/09066a.pdf

20 Michael Nutter, *Greenworks 2009,* https://beta.phila.gov/media/20160419140515/2009-greenworks-vision.pdf, 2.

21 Philadelphia Land Bank, *Strategic Plan and Performance Report 2017.*

22 We recruited interviewees first through a Philadelphia urban farming listserv and subsequently through a network-based approach with our interviewees. Our recruitment targeted urban farming supporters, including gardeners, employees of nonprofit urban greening organizations, and government officials, as well as others who were particularly influential in advocating for policy change. Our interviewees were about 50 per cent male and 50 per cent female and included people who had gardened in Philadelphia for less than a year up to people who had been involved for multiple decades. Some growers had founded their farms and gardens, while others worked or volunteered for farms and gardens run by other non-profit organizations. Since we were focusing on the role this "new wave" of urban growers were playing in setting the policy

agenda in Philadelphia, many of our interviewees were in their twenties and thirties and college-educated, and around 80 per cent were white. Based on our discussions with interviewees, these demographics represented the urban farming demographic that was most vocal at the city level promoting policy change. Several interviewees who were active in urban policy-making around urban agriculture recognized that the people who were most vocal about urban agriculture in policy circles did not reflect the demographics of the community of gardeners across Philadelphia, which is more racially and ethnically diverse. We wanted to explore this dynamic. Many urban agriculture advocates recognized this important underrepresentation, and even over the course of the several years we were working on the book we noticed a shift in the way race and class and urban agriculture is talked about in Philadelphia. In our later interviews we specifically reached out to African American leaders in urban agriculture and talked with them in more depth about how issues of race and class were and were not being addressed. Over the course of the research, the players in policy-making also changed. By 2016, more African American urban agriculture activists were increasingly in leadership roles in city policy-making and active in the framing of urban agriculture in the city. It has been interesting to watch how urban agriculture is now more embedded in a larger conversation about social, economic, and racial justice, and there is a growing recognition of the need for black-led community empowerment and leadership around urban agriculture.

Chapter Two

1 "Voilà! Pop Up Garden Quick Fix for Vacant Lot," *Grid Magazine*, 14 June 2011.
2 George D. McDowell, "Back to the War Gardens," *Philadelphia Evening Bulletin*, 13 April 1932, newspaper clipping collection, Temple University Libraries, Urban Archives, Philadelphia[hereafter TULUA].
3 Philadelphia Vacant Lots Cultivation Association, "Successful Vacant Lots, Season of 1912," The Historical Society of Pennsylvania.
4 Philadelphia Vacant Lots Cultivation Committee, Minutes of the Executive Committee, 6 March 1897.
5 McDowell, "Free Lot Gardeners to Yield $75,000: Season Successful in Spite of Absence of Daylight Savings Law: Pinch of High Prices Increases Movement to Cultivate Vacant Ground," *Philadelphia Evening Bulletin*, 20 August 1920, TULUA.

6 "Potato Patches Will be Doubled: The Vacant Lots Cultivation Committee Has Renewed Its Efforts," *Philadelphia Inquirer*, 15 March 1898, TULUA.

7 Philadelphia Vacant Lots Cultivation Association, "Successful Vacant Lots, Season of 1912."

8 Philadelphia Vacant Lots Cultivation Committee, Minutes of the Executive Committee, 14 November 1900; 2 November 1903. The Historical Society of Pennsylvania.

9 Ibid., 14 May 1900.

10 Minutes of the Executive Committee, Philadelphia Vacant Lots Cultivation Committee, 21 May 1902. The Historical Society of Pennsylvania.

11 Philadelphia Vacant Lots Cultivation Association, "Successful Vacant Lots, Season of 1912."

12 McDowell, "Free Lot Gardeners to Yield $75,000." See also McDowell, "2,000 War Gardens Planned: Vacant Lots Cultivation Association Aims to Double Word This Season; S.S. Fels Renamed President," *Philadelphia Evening Bulletin*, 14 February 1998, TULUA..

13 McDowell, "Vacant Lots Help Many: Report Shows Crops Worth $55,000 Raised Last Year," *Philadelphia Evening Bulletin*, 17 February 1922, TULUA.

14 "Lot Gardens Urged for Relief of Poor: Hundreds of Acres in City Could Be Tilled by Needy, Says Cultivation Association Report," *Philadelphia Evening Bulletin*, 20 January 1915, TULUA.

15 McDowell, "City Dwellers Eager to Farm Vacant Lots: Association Swamped with Applications for Tiny Gardens: Cost of Food Spurs Plan: Appeal to Land Owners to Turn Idle Plots Over for Cultivation," *Philadelphia Evening Bulletin*, 4 March 1917, TULUA.

16 McDowell, "Vacant Lots Help Many: Report Shows Crops Worth $55,000 Raised Last Year, Feb 17, 1922; Vacant Lot Gardens Boon to Many People: Three Hundred New Families Given a Chance in the Awards this Year. Get Season's Supplies: Financial Value of Products Raised May be Put at Close to $100,000," *Philadelphia Evening Bulletin*, 8 August 1920, TULUA.

17 McDowell, "Lots Gardens Urged For Relief of Poor: Hundreds of Acres in City Could Be Tilled by Needy, Says Cultivation Association Report," *Philadelphia Evening Bulletin*, 20 January 1915, TULUA.

18 Philadelphia Vacant Lots Cultivation Association, "Successful Vacant Lots, Season of 1912."

19 McDowell, "Vacant Lot Gardens Boon to Many People."

20 McDowell, "Now My Idea Is This! Daily Talks with Thinking Philadelphians on Subjects They Know Best: John K. Snyder: On the World of the Philadelphia Vacant Lots Cultivation Association," *Philadelphia Evening Bulletin*, 24 October 1924, TULUA.

21 "Potato Patches Will Be Doubled."
22 Philadelphia Vacant Lots Cultivation Association, "Successful Vacant Lots, Season of 1912."
23 McDowell, "Now My Idea Is This!"
24 Ibid.
25 "Potato Patches Will Be Doubled."
26 Philadelphia Vacant Lots Cultivation Association, "Successful Vacant Lots, Season of 1912."
27 "Govt. Ban Lifted; Building Plans on Spring Programme Can be Pushed without Waste, Says F," *Philadelphia Inquirer*, 1 December 1918.
28 "Construction Is Key to Revival of Trade: When Building Starts on Huge Scale Demanded, All Industry Will Benefit," *Philadelphia Inquirer*, 25 April 1921.
29 "West Market St. Undergoing Boom: Thoroughfare between Fifteenth and Schuylkill Changes Vastly in Decade," *Philadelphia Inquirer*, 23 May 1921.
30 McDowell, "Now My Idea Is This!"
31 McDowell, "Building Boom is Blow to Vacant Lot Gardens," *Philadelphia Evening Bulletin*, 22 March 1924, TULUA.
32 Philadelphia Vacant Lots Cultivation Committee, Minutes of the Executive Committee, 10 November 1927.
33 Ibid., 27 September 1928.
34 Samuel Fels, PVLCA president, Letter to Friends of PVLCA, from Samuel Fels, President PVLCA, 3 December 3, 1928.
35 Diane Ying, "Two Help Put Bloom onto Concrete Cities," *Philadelphia Inquirer*, 7 August 1969.
36 McDowell, "Garden Club Dream Comes True," *Philadelphia Evening Bulletin*, 2 April 1963, TULUA.
37 McDowell, "Flower Box Called Symbol of Leadership," *Philadelphia Evening Bulletin*, 17 April 1963, TULUA.
38 Ying, "Two Help Put Bloom on Concrete in Cities," *Philadelphia Evening Bulletin*, 7 August 1969, TULUA.
39 Eugene L. Meyer, "100,000 Plant Flowers at Homes: Other Cities Copy and Praise Idea," *Philadelphia Evening Bulletin*, 20 March 1966, TULUA.
40 Howard Shapiro, "Flower Power: How the Horticultural Society Cultivates Neighborhoods and Community Leaders – and Exercises Growing Responsibility for the City's Future." *Inquirer*, 23 February 2003.
41 Patricia Spollen, "Philadelphia Parks to Get 650 Trees in Fall for Bicentennial," *Philadelphia Evening Bulletin*, 4 October 1974, TULUA.
42 Mary-Virginia Geyelin, "Gardenmobile Plans go Rolling Along," *Philadelphia Evening Bulletin*, 29 January 1976, TULUA.

43 Spollen, "Neighbors Create 6 'Sitting Parks,'" *Philadelphia Evening Bulletin*, 25 June 1978, TULUA.
44 Samuel Fitch Hotchin, " Penn's Greene Country Towne: pen and pencil sketches of early Philadelphia and its prominent characters," 1903. Ferris and Leach. Princeton University.
45 Neighborhood Gardens Trust, "Frequently Asked Questions." http://www.ngtrust.org/index.php/faq.

Chapter Three

1 Philadelphia City Planning Commission, *Vacant Land in Philadelphia: A Report on Vacant Land Management and Neighborhood Restructuring* (1995).
2 Laura Blanchard, 2011, *Garden in the Dumpster*. South Philly Blocks 2005. http://www.southphillyblocks.org/photos_essays/2100fitzwater/index.html#updates.
3 Neighborhood Gardens Trust, "Frequently Asked Questions," http://www.ngtrust.org/index.php/faq.
4 US Department of Agriculture, Economic Research Service, report to Congress: *Access to Affordable and Nutritious Food: Measuring and Understanding Food Deserts and Their Consequences. A Report to Congress* (June 2009), https://www.ers.usda.gov/webdocs/publications/ap036/12716_ap036_1_.pdf.

Chapter Four

1 City of Philadelphia, Philadelphia Water Department, "Green City, Clean Waters: The City of Philadelphia's Program for Combined Sewer Overflow Control: A Long Term Control Plan Update" (September 2009).
2 Stratus Consulting, "A Triple Bottom Line Assessment of Traditional and Green Infrastructure Options for Controlling CSO Events in Philadelphia's Watersheds" (Final Report), 24 August 2009, https://www.epa.gov/sites/production/files/2015-10/documents/gi_philadelphia_bottomline.pdf.
3 Nicole Clark, "A Farm Grows in Somerton: Non-Profit Brings Agriculture Back to the Northeast," *Philadelphia Inquirer*, 21 May 2003.
4 Some of the supporters of the project include 1957 Charity Trust, Allen Hilles Fund, Dolfinger-McMahon Foundation, Douty Foundation, Pennsylvania State Senator Michael J. Stack, Keystone Development Center, Philadelphia Commerce Department, Philadelphia Workforce Development Corporation, the Pennsylvania Department of Agriculture,

Pennsylvania Department of Environmental Protection, and the US
Department of Agriculture – Natural Resources Conservation Services.
Visit Somerton Tanks Farm website, http://www.somertontanksfarm.org.
5 For more information about the history of this site, see Anne Spirn's
writings for the West Philadelphia Landscape Project at http://www
.millcreekurbanfarm.org/content/frequently-asked-questions. Spirn has
written extensively about the subsidence issues in the Mill Creek area.
6 Jade Walker with Suzy Subways, "In Defense of the Land: The Mill
Creek Farm and Brown Street Community Garden," https://
millcreekurbanfarm.org/sites/.../In%20Defense%20of%20the%20Land
.doc. Note that this document does not have a date.
7 Council of the City of Philadelphia Committee on Housing, Neighborhood Development, and the Homeless [hereafter PCH], 16 June 2010, 117,
http://legislation.phila.gov/transcripts/Public%20Hearings/housing/
2010/ho061610.pdf.
8 Jade Walker with Suzy Subways, "In Defense of the Land: The Mill Creek
Farm and Brown Street Community Garden," https://millcreekurbanfarm
.org/sites/.../In%20Defense%20of%20the%20Land.doc
9 PCH, 16 June 2010, 53–4, http://legislation.phila.gov/transcripts/
Public%20Hearings/housing/2010/ho061610.pdf.
10 According the to Pew Charitable Trusts July 2015 report, *Philadelphia's
Councilmanic Prerogative: How It Works and Why It Matters*, councilmanic
prerogative shapes land use in the city because "in Philadelphia, the vast
majority of land use decisions, small or momentous, are made individually by City Council's 10 members, who represent geographical districts
across the city. (The remaining seven members are elected at large.) The
practice, which is grounded in legislative tradition rather than law, is
known as 'councilmanic prerogative.' It comes into play largely in the
disposition of city-owned properties and in zoning matters, regardless
of whether those decisions formally come before City Council as a
whole"(1). http://www.pewtrusts.org/~/media/assets/2015/08/
philadelphia-councilmanic-report--with-disclaimer.pdf?la=en.
11 PCH, 16 June 2010, 53–4, http://legislation.phila.gov/transcripts/
Public%20Hearings/housing/2010/ho061610.pdf.
12 See the 2010 documentary film, *West Philly Grown* by Clay Hereth,
https://vimeo.com/23669004.
13 "Saint Bernard Community Garden," https://stbernardgarden
.wordpress.com/2012/12/19/garden-saved-at-tense-sheriffs-sale.
14 As quoted in Max Marin, "After Dodging Sheriff's Sale, Community
Gardens Look Ahead to the Land Bank," *Plan Philly*, 25 May 2016,

http://planphilly.com/articles/2016/05/25/after-dodging-sheriff-s-sale-community-gardens-look-ahead-to-the-land-bank.

15 Pew Charitable Trusts, *Philadelphia's Councilmanic Prerogative*, 1, 22.

16 City of Philadelphia, "Philadelphia Food Charter," https://phillyfood justice.files.wordpress.com/2011/06/philadelphia_food_charter.pdf.

17 Michael Nutter, *Greenworks 2009*, https://beta.phila.gov/ media/20160419140515/2009-greenworks-vision.pdf, 6.

18 Ibid.

19 Walker, Keane, and Burke, "Disparities and Access to Healthy Food."

20 Here it is important to note the impact that councilmanic prerogative has on planning and politics in Philadelphia. Council members have an enormous amount of say over what happens in their districts.

21 Philadelphia Food Charter, https://phillyfoodjustice.files.wordpress .com/2011/06/philadelphia_food_charter.pdf.

22 For an up-to-date list of FPAC members, visit http://phillyfpac.org/ about/

23 Food Policy Advisory Council, "Subcommittees," https://phillyfpac .org/subcommittees.

24 Here is a list of the FPAC appointed members: https://phillyfpac.org/ appointedmembers.

25 "Soil Generation!," http://groundedinphilly.org/HFGS-about.

26 Ibid.

27 "Grounded in Philly." http://www.groundedinphilly.org/blog/2016/ 06/30/city-official-are-digging-urban-agriculture-philly-read-amazing-work-moving-forward; http://www.pubintlaw.org/wp-content/ uploads/2015/10/land_toolkit_webrev.pdf

28 City of Philadelphia Request for Information, 10 March 2010, www .rfpdb.com/process/download/name/Chemical-Free-Farming.pdf.

29 Shannon Simcox, "Neighbors Line Up to Block Farm Plan," *Chestnut Hill Local*, 29 July 2010. Web.

30 City of Philadelphia, Bill no. 100498, http://legislation.phila.gov/ attachments/10743.pdf; "Philadelphia Decoded," http://phillycode .org/14-505.

31 As quoted in Miriam Hill, "Philadelphia Council Panel Opposes Expanded Community Garden at Manatawna Farm," *Philadelphia Inquirer,* 21 October 2010, http://www.cityfarmer.info/2010/10/21/ philadelphia-council-panel-opposes-expanded-community-garden-at-manatawna-farm.

32 For more information about the Green 2015 planning process, see http:// planphilly.com/green2015.

33 "Greenworks Philadelphia: A Blog by the Office of Sustainability,"
   https://greenworksphila.wordpress.com/2010/12/.
34 Penn Praxis, "An Action Plan for 500 New Acres: 2010," https://www
   .design.upenn.edu/pennpraxis/work/
   green-2015-action-plan-500-new-acres.
35 Food Policy Advisory Council, Minutes, http://phillyfpac.files
   .wordpress.com/2013/01/12-4-meeting-minutes.pdf.
36 City of Philadelphia, "Parks and Recreation: Urban Agriculture," http://
   www.phila.gov/ParksandRecreation/environment/Pages/
   UrbanAgriculture.aspx.
37 Ibid.
38 Heather A. Davis, "Farm and Resource Center Connects Community
   with Land, Food, and One Another," *Penn Current*, 4 April 2015, https://
   penncurrent.upenn.edu/2015-04-16/features/farm-and-resource-center-
   connects-community-land-food-and-one-another.
39 Nutter, 2009, 54.
40 Cary Betagole, "In Brewerytown Pushing to Revive an Urban Farm,"
   Hidden City, December 13, 2013, http://www.brewerytowngarden.com/
   blog/throwback-thursday.
41 Mike Dent, "Whatever Happened with Marathon Grill's Farm?
   Brewerytown Garden Has Risen from Its Weeds," *BillyPenn.com*, https://
   billypenn.com/2015/07/22/
   whatever-happened-with-marathon-grills-farm-brewerytown-garden-
   has-risen-from-its-weeds.
42 Dan Geringer, "Brewerytown Garden Rises Marathon Farm Ruins,"
   Philly.com, 5 August 2015, http://articles.philly.com/2015-08-05/
   entertainment/65208969_1_gardeners-weeds-redevelopment-authority.
43 City of Philadelphia, "Citywide Vision: Philadelphia 2035" (2011),
   http://phila2035.org/pdfs/final2035vision.pdf.
44 Ibid, 71.
45 "Growing Food Connections, The Philadelphia Code, Title 14," http://
   growingfoodconnections.org/gfc-policy/
   the-philadelphia-code-title-14-zoning-and-planning-bill-no-110845.
46 City of Philadelphia, "Philadelphia 2035: Planning and Zoning for a
   Healthier City: 2010," http://phila2035.org/wp-content/uploads/
   2011/02/Phila2035_Healthier_City_Report.pdf, 23.
47 Jon McGoran, "Zoning Amendment Threatens Urban Farms in Philly,"
   GRID Magazine, 18 January 2013, http://www.gridphilly.com/grid-
   magazine/2013/1/18/zoning-amendment-threatens-urban-farms-in-
   philly.html. For a copy of the proposed amendment visit http://

planphilly.com/uploads/media_items/bill-no-12091701-as-amended.
original.pdf.

48 Christine Fisher, "Councilman O'Neill's Amendments Hit Community
Gardens," *Plan Philly*, 17 January 2013, http://planphilly.com/
eyesonthestreet/2013/01/17/
councilman-o-neill-s-amendments-hit-community-gardens.

49 Amy Laura Cahn as quoted in the City Council Committee on Rules
Hearing of 4 December 2012, Testimony of the Public Interest Law Center
of Philadelphia, Bills no. 120916 and no. 120917, "Public Interest Law
Center of Philadelphia", http://www.pilcop.org/wp-content/uploads/
2012/11/PILCOP-Testimony-on-Zoning-Amendments.pdf.

50 Public Interest Law Center of Philadelphia, "Garden Justice Legal
Initiative," http://www.pilcop.org/take-action-to-protect-urban-
agriculture-in-philadelphia/#sthash.aBoZ5A9P.dpbs. According to the
website, Healthy Food, Green Spaces has since changed its name to Soil
Generation!, whose mission is to "support equity and social justice for
community-managed green space, gardens, and farms through advocacy,
grassroots organizing, and community education." More information
about Soil Generation! can be found at http://www.groundedinphilly
.org/HFGS-about.

51 For more information on Detroit, see James Trimarco, "Detroiters
Question 'World's Largest Farm'", *Yes!Magazine*, 11 December 2012,
http://www.yesmagazine.org/peace-justice/detroit-community-
gardeners-question-hantz-urban-farm.

52 Econsult Corporation, "May 8 Consulting and Penn Institute for Urban
Research, Vacant Land Management in Philadelphia: The Costs of the
Current System and the Benefits of Reform"; Philadelphia Redevelopment Authority and Philadelphia Association of CDCs (November 2010).
http://may8consulting.com/pub_16.html (RDA Report), 1.

53 Comments to Draft City Disposition Policy Feedback from PUFN and
FORC meetings and synthesis of best practices, Johanna Rosen, Mill
Creek Farm and FPAC member Amy Laura Cahn, Garden Justice Legal
Initiative, 22 December 2011, http://www.pilcop.org/wp-content/
uploads/2012/04/PUFN_FORC.pdf.

54 Food ORganizing Collaborative, https://phillyforc.wordpress.com.

55 Ibid.

56 Food ORganizing Collaborative, 26 January 2012, letter to John Carpenter
and Bridget Collins-Greenwald, http://www.pilcop.org/wp-content/
uploads/2012/04/FORC_LETTER.pdf.

57 Ibid.

58 The latest disposition policy for gardens can be found at Philadelphia Land Bank, *2017 Strategic Plan and Performance Report*, http://www .philadelphialandbank.org/wordpress/wp-content/uploads/2015/01/ philadelphia-land-bank-strategic-plan-february-2017.pdf, 67.

59 Philadelphia Redevelopment Authority, "Policies for the Sale and Reuse of City Owned Property," 30 October 2014, 13, https://www .philadelphiaredevelopmentauthority.org/sites/default/files/Land%20 Disposition%20Policy%20October%2030%202014_0.pdf.

60 Ibid., 14.

61 Ibid., 14.

62 City of Philadelphia, "Policies for the Sale and Reuse of City Owned Property, Approved by Philadelphia City Council," 11 December 2014, http://www.philadelphialandbank.org/wordpress/wp-content/ uploads/2015/02/Final-Disposition-Policy-December-2014.pdf, 13-15; "Philadelphia Redevelopment Authority," Frequently Asked Questions about Buying Property, https://www.philadelphiaredevelopmentau-thority.org/sites/default/files/Frequently%20Asked%20Questions-plw.pdf . It is worth noting that the range of time for the leases of community gardens varies in different city documents. The Philadelphia Land Bank website, http://www.philadelphialandbank .org/i-am-a, says that a Community Garden Agreement can be for up to five years; however, in the *2017 Philadelphia Land Bank Strategic Plan and Performance Report*, http://www.philadelphialandbank.org/ wordpress/wp-content/uploads/2015/01/Land-Bank-Strategic-Plan-to-Council-01-20-17.pdf, 67, it says, "Community gardens and community-managed Open Spaces are eligible for leases for up to one year in length, when approved by the Land Bank Board of Directors. In some instances where a proposed lessee seeks a term in excess of one year, the application must be approved by the Vacant Property Review Committee and City Council."

63 Food Policy Advisory Council, FPAC Minutes from 11 February 2013 meeting, http://phillyfpac.files.wordpress.com/2013/03/fpac-general-meeting-minutes-2-11-13.pdf, 5.

64 Philadelphia Land Bank, Board Meeting Minutes, 12 May 2016, http:// www.philadelphialandbank.org/wordpress/wp-content/ uploads/2016/06/Minutes-May-12-2016.pdf.

65 Econsult Corporation, Penn Institute for Urban Research, and May 8 Consulting, "Vacant Land Management in Philadelphia: The Cost of the Current System and the Benefits of Reform," Executive Summary, November 2010," http://planphilly.com/uploads/media_items/

http-planphilly-com-sites-planphilly-com-files-econsult_vacant_land_
full_report-pdf.original.pdf

66 Ibid., 11.
67 John Kromer, "A Philadelphia Land Bank: Why We Need to Get It Right,"
    Plan Philly, 23 March 2012; http://planphilly.com/
    eyesonthestreet/2012/03/23/a-philadelphia-land-bank-why-we-
    need-to-get-it-right-dagspace; "The Philadelphia Land Bank website
    FAQ," http://www.philadelphialandbank.org/faqs.
68 Philadelphia Land Bank, "About Us," http://www.philadelphialand
    bank.org/about/
69 For more information on the Philly Land Bank Alliance, visit http://
    www.phillylandbank.org/philly-land-bank-alliance.
70 "Campaign to Take Back Vacant Land," http://takebackvacantland.org/
    ?page_id=122.
71 Philadelphia Land Bank, "2015 Land Bank Strategic Plan and Disposition
    Policies," 159, http://www.philadelphialandbank.org/about/
    strategic-plan.
72 Ibid., 141.
73 Philadelphia Land Bank, *2017 Strategic Plan and Performance Report*, http://
    www.philadelphialandbank.org/wordpress/wp-content/uploads/2015/
    01/philadelphia-land-bank-strategic-plan-february-2017.pdf, 67.
74 Neighborhood Gardens Trust, "Priority Acquisition Plan," https://
    phsonline.org/uploads/resources/NGT_PriorityAcquisitionPlan_
    lowres.pdf.
75 "The Philadelphia Land Bank," http://www.philadelphialandbank.org/
    land-bank-to-partner-with-ngt.
76 Neighborhood Gardens Trust, "Frequently Asked Questions," http://
    www.ngtrust.org/index.php/faq.
77 Food Policy Advisory Council, *FPAC Annual Report 2016*, https://
    phillyfpac.files.wordpress.com/2016/08/annual-report-2016-final.pdf,
    12.
78 Ibid.
79 Amy Laura Cahn as quoted in Samantha Melamed, "South Kensington
    Gardeners Take on Developers in Court," *Philly.com*, 18 March 2016,
    http://www.philly.com/philly/living/South_Kensington_gardeners_
    take_on_developer_in_court.html#UbXlegPwt6Zq3hgG.99.
80 Catalina Jaramillo, "Neighborhood Gardens Trust Targets Preservation
    for 28 Gardens," *Plan Philly*, 12 October 2016, http://planphilly.com/
    articles/2016/10/13/neighborhood-gardens-trust-targets-preservation-
    for-28-more-gardens.

81 Ibid.
82 Jamillah Meekins's testimony at the 21 September, 2016 Philadelphia City Council hearings on urban agriculture, http://legislation.phila.gov/ transcripts/Public%20Hearings/environment/2016/en092116.pdf, 76–9.
83 Posted on *Grounded in Philly* (blog) by Owen Taylor on 22 August 2013, http://groundedinphilly.org/blog/?page=6.
84 Council president Darrell Clarke raised concerns about the fact that the Revenue Department had put up land for sheriff sale. He argued that they were "bypassing Land Bank review for strategic or community-oriented reuse." See Clarke, "Clarke: Time for Moratorium on Sales of Vacant, Tax-Delinquent Land," *Philly.com*, 23 January 2017, http://www.philly.com/philly/opinion/20170123_Clarke__Time_for_moratorium_on_sales_of_vacant__tax-delinquent_land.html.
85 Linda Walton, Petition to Councilmember Kenyatta Johnson, "Save the Cleveland Street Neighborhood Garden from Development," on *Change.org*.
86 Ibid.
87 Both authors participated in the FPAC Soil Safety Workshops.
88 Catalina Jaramillo, "Free Soil Testing for City-Owned Gardens and Lots through Brownfields Grant," *Plan Philly*, 3 October 2016, http://planphilly.com/articles/2016/10/03/free-soil-testing-for-city-owned-gardens-and-lots-through-brownfields-grant.
89 Ibid.
90 Philadelphia Water, "PWD Urban Gardens and Farms," http://www.phila.gov/water/PDF/UrbanGardenersFactSheet.pdf.
91 Public Interest Law Center of Philadelphia, Testimony on Stormwater Billing Exemption for Gardens and Farms, http://www.pilcop.org/testimony-on-stormwater-billing-exemption-for-gardens-and-farms/.
92 As quoted in Haywood Brewster, "Councilwoman Sánchez Introduces Stormwater Fee Exemption For Community Gardens," *Philadelphia Free Press*, 25 May 2016, http://weeklypress.com/councilwoman-snchez-introduces-stormwater-fee-exemption-for-community-gard-p6089-1.htm.
93 Chris Hepp, "Philadelphia Ballot Issues Easily Pass," *Philadelphia Inquirer*, 5 November 2014, http://www.philly.com/philly/news/politics/elections/20141105_Phila_ballot_issues_-_including_creating_an__Office_of_Sustainability_-_easily_pass.html.
94 Katrina Baxter, "Resolution Calling for Hearings on Expanding Urban Farming in Philadelphia," *GroundedinPhilly.com*, 3 May 2016, http://www.groundedinphilly.org/blog/2016/05/03/resolution-calling-hearings-expanding-urban-farming-philadelphia.

95 Ibid.
96 Tom MacDonald, "Philly City Council Cultivating Interest in Vertical Farming," *Newsworks*, 26 April 2016, http://www.newsworks.org/index .php/local/philadelphia/93245-philly-city-council-cultivating-interest-in-vertical-farming.
97 Kirtrina Baxter, "City Officials are 'Digging In' to Urban Agriculture in Philly" (blog), http://www.groundedinphilly.org/blog/2016/06/30/ city-official-are-digging-urban-agriculture-philly-read-amazing-work-moving-forward,
98 Food Policy Advisory Council, *FPAC Annual Report 2016*, https:// phillyfpac.files.wordpress.com/2016/08/annual-report-2016-final.pdf, 12.
99 Amy Laura Cahn, "It's Time to Double Down on Protecting Community Gardens," *Grid Magazine*, 31 May 2016, http://www.gridphilly.com/ grid-magazine/2016/3/28/its-time-to-double-down-on-protecting-community-gardens.

Chapter Five

1 Meenar, Mahbubur R., and Brandon M. Hoover. "Community food security via urban agriculture: Understanding people, place, economy, and accessibility from a food justice perspective." *Journal of Agriculture, Food Systems, and Community Development* 3, no. 1 (2012): 143-160, 149.
2 "Soil Generation!," http://groundedinphilly.org/HFGS-about.
3 Adrian Higgins, "Whether White House Garden Keeps Growing, Michelle Obama Has Planted Seed of Healthy Eating," *Washington Post*, 12 October 2016, https://www.washingtonpost.com/lifestyle/whether-white-house-garden-keeps-growing-michelle-obama-has-planted-seed-of-healthy-eating/2016/10/14/ d3967ae0-91c0-11e6-a6a3-d50061aa9fae_story.html?tid=a_inl&utm_ term=.150409046e1c; National Farm to School Network, http://www .farmtoschool.org; US Department of Agriculture, Food and Nutrition Service, "Community Food Systems", https://www.fns.usda.gov/ farmtoschool/farm-school.
4 "Stories of Success," The Merchants Fund. 2013, http://www .merchantsfund.org/wp-content/uploads/2010/11/TMF_Case-Studies_ 12102013_vF.pdf, 11–12.
5 Inhabitat, "Farmadelphia," http://inhabitat.com/farmadelphia.
6 Delaware Valley Regional Planning Commission, "Greater Philadelphia Food Systems Study," January 2010, http://www.dvrpc.org/reports/ 09066a.pdf.

7 Nick Carey, "Dan Gilbert and Quicken Loans Invest in Detroit Real Estate, Bet on Downtown Urban Center," *Huffington Post*, 19 February 2013, http://www.huffingtonpost.com/2013/02/19/den-gilbert-quicken-loans-detroit-real-estate_n_2715609.html.

8 Note that many of these interviews were conducted when the city was still in an economic recession.

9 According to the Bureau of Labor Statistics, the July 2014 Philadelphia County unemployment rate was 7.8 per cent. Visit http://www.bls.gov/ro3/urphl.htm.

10 County Health Rankings and Road Map, http://www.countyhealth rankings.org/app/pennsylvania/2012/rankings/philadelphia/county/outcomes/overall/snapshot.

11 Our own experience with some of our students graduating from Temple University has been that they want to stay in Philadelphia to be urban farmers or to be involved in some way in urban agriculture or urban greening. Some of our former students have started farms and gardens in the city since graduating.

12 Lee Stabert, "Cost of Living: How Expensive is Philadelphia?" *Keystone Edge*, 3 November 2011, http://www.keystoneedge.com/2011/11/03/cost-of-living-how-expensive-is-philadelphia.

13 Forbes Staff, "America's Most Expensive Cities," *Forbes.com*, 7 October 2009, http://www.forbes.com/2009/10/07/most-expensive-cities-lifestyle-real-estate-america-top-ten.html

14 Pennsylvania Horticultural Society, "PHS City Harvest," https://phsonline.org/programs/city-harvest.

15 Here it is also interesting to note the generational divide we found between the older, non-white community gardeners and farmers, who were not usually a part of these listservs and networks, and the younger set of urban farmers and gardeners, who relied heavily on them for everything from worms, seeds, compost, to legal advice, and so on. While it was not originally a focus of our research, we started to notice the effects of the digital divide in urban agriculture. Not being connected to the Internet might mean missed opportunities to network and connect to financial resources such as grants. It also might mean not being connected to the ongoing political conversations about how to reform the City's zoning laws so that they would promote urban agriculture. At the same time, we found that some of the more established African American and Puerto Rican gardeners were growing in gardens that were already protected. PHS and the Neighborhood Gardens Association have worked with some of the older African American gardens to ensure that their

land is in a land trust. Since these gardens in the land trust have been around a lot longer than the newer gardens and their land title is secure, many of the conversations held on PUFN that focus on guerilla gardening and gaining access to land might not be as relevant.

16 *Grid Magazine*, a free magazine focusing on Philadelphia's sustainability movement, also often highlights the work of urban gardeners and farmers and focuses on the local food movement.

17 This was our impression after talking with older African American farmers at several gardens around the city.

18 Sarah Dooling, "Ecological Gentrification: A Research Agenda Exploring Justice in the City," *International Journal of Urban and Regional Research* 33, no. 3 (2009): 621–39; N. Quastel, "Political Ecologies of Gentrification, *Urban Geography* 30(7) (2009):, 694–725; H. Pearsall, "From Brown to Green? Assessing Social Vulnerability to Environmental Gentrification in New York City," *Environment and Planning C: Government and Policy* 28(5) (2010): 872–86; K. Reynolds, "Disparity Despite Diversity: Social Injustice in New York City's Urban Agriculture System," *Antipode*, 47(1) (2015): 240–59; J. Wolch, J. Byrne, and J. Nowell, "Urban Green Space, Public Health, and Environmental Justice: The Challenge of Making Cities 'Just Green Enough,'" *Landscape and Urban Planning* 125 (2014): 234–44.

Chapter Six

1 Christopher Bolden-Newsome, testimony at Philadelphia City Council hearing of the Committee on the Environment, City of Philadelphia, 21 September 2016, 104–5.

2 For more information, visit Greensgrow's website http://www .greensgrow.org.

3 Neighborhood Gardens Trust, "Priority Acquisition Plan," March 2016, https://phsonline.org/uploads/resources/NGT_ PriorityAcquisitionPlan_lowres.pdf.

4 The Food Trust, "The Food Trust: What We Do," http://thefoodtrust .org/what-we-do/corner-store.

5 The Food Trust, "The Food Trust's Farmer's Markets," http:// thefoodtrust.org/farmers-markets.

6 The Food Trust, "Philly Food Bucks," http://thefoodtrust.org/ phillyfoodbucks.

7 Farm to City, "Farmer's Markets," https://farmtocity.org/find-local-food/farmers-markets.

8 Food Policy Advisory Council, "Soil Safety and Urban Gardening in Philadelphia" (2015), https://phillyfpac.files.wordpress.com/2015/04/soil-safety-process-and-recommendations-report.pdf.

Chapter Seven

1 Michael Nutter, *Greenworks 2009*, 2, https://beta.phila.gov/media/20160419140515/2009-greenworks-vision.pdf.
2 Ibid.
3 Ibid.
4 Ibid., 6.
5 For a video of the City Council hearing, see https://www.youtube.com/watch?v=nafY_Gl6FkQ.
6 Philadelphia Water Department, 1 June 2011, *Green City, Clean Waters*, http://www.phillywatersheds.org/doc/GCCW_AmendedJune2011_LOWRES-web.pdf, *Green City, Clean Waters* uses a triple bottom line analysis.

Works Cited

Alaimo, Katherine, Elizabeth Packnett, Richard A. Miles, and and Daniel J. Kruger. 2008. "Fruit and Vegetable Intake among Urban Community Gardeners." *Journal of Nutrition Education and Behavior* 40(2): 94–101. https://doi.org/10.1016/j.jneb.2006.12.003.

Alkon, Alison Hope, and Julian Agyeman. 2011. *Cultivating Food Justice: Race, Class, and Sustainability*. Cambridge, MA: MIT Press.

Armstrong, Donna. 2000. "A Survey of Community Gardens in Upstate New York: Implications for Health Promotion and Community Development." *Health and Place* 6(4): 319–27. https://doi.org/10.1016/S1353-8292(00)00013-7.

Bassett, Thomas. 1979. *Vacant Lot Cultivation: Community Gardening in America, 1893–1978, Geography*. Berkeley: University of California Press.

Bedore, Melanie. 2010. "Just Urban Food Systems: A New Direction for Food Access and Urban Social Justice." *Geography Compass* 4(9): 1418–32. https://doi.org/10.1111/j.1749-8198.2010.00383.x.

Bentley, A. 1998. *Eating for Victory: Food Rationing and the Politics of Domesticity*. Chicago: University of Illinois Press.

Blair, Dorothy. 2009. "The Child in the Garden: An Evaluative Review of the Benefits of School Gardening." *Journal of Environmental Education* 40(2): 15–38. https://doi.org/10.3200/JOEE.40.2.15-38.

Blake, Harriet. 1981. "Looking for Growing Space in the City." *Washington Post*, 14 June 1981.

Blanchard, Laura. 2011. *Garden in the Dumpster*. South Philly Blocks 2005. http://www.southphillyblocks.org/photos_essays/2100fitzwater/index.html#updates.

Bonham, J. Blaine, Gerri Spilka, and Darl Rastorfer. 2002. "Old Cities/Green Cities." Planning advisory service report number 506/507. Chicago, IL.

Bush-Brown, Louise Carter. 1969. *Garden Blocks for Urban America*. New York: Scribner.

Campbell, Douglas. 1998. "Teens Develop Garden – and Character." *Philadelphia Inquirer*, 23 April.

Cowie, Denise. 1999. "Changing the City's Colors Twenty-Five Years Ago, the Seeds of Philadelphia Green Were Planted, They Flourished, Helping Neighborhoods Turn Dreary Landscapes into Things of Beauty." *Philadelphia Inquirer*, 24 September.

– 2001. "A Church Partnership Allows N. Philadelphia Children to Tend Their Own Vegetable Plot – and Learn Something About Self-Reliance." *Philadelphia Inquirer*, 3 August.

D'Abundo, Michelle L., and Andrea M. Carden. 2008. "Growing Wellness': The Possibility of Promoting Collective Wellness through Community Garden Education Programs." *Community Development* 39(4): 83–94. https://doi.org/10.1080/15575330809489660.

Darling, Heather. 1983. "Re-Greening of Philadelphia." *Christian Science Monitor*, 22 July.

Delaware Valley Regional Planning Commission. 2010. "Greater Philadelphia Food System Study." http://www.dvrpc.org/reports/09066a.pdf.

Dooling, Sarah. 2009. "Ecological Gentrification: A Research Agenda Exploring Justice in the City." *International Journal of Urban and Regional Research* 33, no. 3: 621–39.

Draper, Carrie, and Darcy Freedman. 2010. "Review and Analysis of the Benefits, Purposes, and Motivations Associated with Community Gardening in the United States." *Journal of Community Practice* 18(4): 458–92. https://doi.org/10.1080/10705422.2010.519682.

Ferris, John, Carol Norman, and Joe Sempik. 2001. "People, Land, and Sustainability: Community Gardens and the Social Dimension of Sustainable Development." *Social Policy and Administration* 35(5): 559–68. https://doi.org/10.1111/1467-9515.t01-1-00253.

Firth, Chris, Damian Maye, and David Pearson. 2011. "Developing 'Community' in Community Gardens." *Local Environment* 16(6): 555–68. https://doi.org/10.1080/13549839.2011.586025.

Fleming, Leonard. 2003. "Horticultural Society Team Up for Cleanup; A One-Year Deal Calls for the Society to Beautify and Maintain 1,000 Pieces of Land throughout the City." *Philadelphia Inquirer*, 23 September.

Florida, Richard L. 2014. *The Rise of the Creative Class Revisited*. New York: Basic Books.

Francke, Caitlin. 2003. "Anti-Blight Effort Turns to Mantua; It Is the Site of the Program's First Large-Scale Razing." *Philadelphia Inquirer*, 9 April.

Glover, Troy D., Kimberly J. Shinew, and Diana C. Parry. 2005. "Association, Sociability, and Civic Culture: The Democratic Effect of Community Gardening." *Leisure Sciences* 27(1): 75–92. https://doi.org/10.1080/01490400590886060.

Gordon, S. 1996. "Sowing Garden of Hope in Chester: Children Learn That Seeds, Even in Boxes on Concrete, Reach for the Sky, So Will Young Minds, If Nurtured: That's the Hope of Urban Gardening Program." *Philadelphia Inquirer*, 10 June.

Gregory, Kia. 2009. "Veggie Kids Bond with Nature, Neighbors " *Philadelphia Inquirer*, 8 September.

Guthman, Julie. 2008. "Bringing Good Food to Others: Investigating the Subjects of Alternative Food Practice." *Cultural Geographies* 15(4): 431–47. https://doi.org/10.1177/1474474008094315.

Harnik, Peter. 2010. *Urban Green: Innovative Parks for Resurgent Cities.* Washington, DC: Island Press.

Heavens, Alan. 1994. "Green Thumbs Up! More and More Urbanites Are Happy to Reap What They Sow." *Philadelphia Inquirer*, 10 June.

Hodgson, Kimberley, Marcia Caton Campbell, and Martin Bailkey. *Urban Agriculture: Growing Healthy, Sustainable Places.* Chicago and Washington, D.C.: American Planning Association, 2011.

Hynes, H. Patricia. 1996. "A Patch of Eden: America's Inner City Gardeners." White River Junction: Chelsea Green Publishing Company.

Irvine, Seana, Lorraine Johnson, and Kim Peters. 1999. "Community Gardens and Sustainable Land Use Planning: A Case Study of the Alex Wilson Community Garden." *Local Environment* 4(1): 33–46. https://doi.org/10.1080/13549839908725579.

Jacobs, Jane. 1961. *The Death and Life of American Cities.* New York: Random House.

Kurtz, Hilda. 2001. "Differentiating Multiple Meanings of Garden and Community." *Urban Geography* 22(7): 656–70. https://doi.org/10.2747/0272-3638.22.7.656.

Lawson, Laura. 2004. "The Planner in the Garden: A Historical View into the Relationship between Planning and Community Gardens." *Journal of Planning History* 3(2): 151–76. https://doi.org/10.1177/1538513204264752.

– 2005. *City Bountiful: A Century of Community Gardening in America.* Berkeley: University of California Press.

Loeb, Vernon. 1994. "L&I Demolitions Boom, Changing the Face of Phila.: Following the Wrecking Ball to a Shift in City Planning Policy." *Philadelphia Inquirer*, 16 May.

Logan, John R., and Harvey Molotch. 1987. *Urban Fortunes: The Political Economy of Place*. Berkeley: University of California Press.

Lovell, Sarah T., and Douglas M. Johnston. 2009. "Designing Landscapes for Performance Based on Emerging Principles in Landscape Ecology." *Ecology and Society* 14(1): 44. https://doi.org/10.5751/ES-02912-140144.

Marder, Diane. 2011. "Young, College-Educated, They Harvest Food and More from Empty Lots, Rooftops." *Philly.com*, 21 July.

McClintock, Nathan. 2014. "Radical, Reformist, and Garden-Variety Neoliberal: Coming to Terms with Urban Agriculture's Contradictions." *Local Environment* 19(2): 147–71. https://doi.org/10.1080/13549839.2012.752797.

Meenar, Mahbubur R., and Brandon M. Hoover. 2012. "Community food security via urban agriculture: Understanding people, place, economy, and accessibility from a food justice perspective." *Journal of Agriculture, Food Systems, and Community Development* 3(1): 143-160.

Miller, C. 2003. "In the Sweat of Our Brow: Citizenship in American Domestic Practice during WWII – Victory Gardens." *Journal of American Culture* 26(3): 395.

Mukherji, Nina, and Alfonso Morales. 2010. "Zoning for Urban Agriculture." In *Zoning Practice*. Chicago: American Planning Association.

Nutter, Michael. 2009. *Greenworks*. Philadelphia: City of Philadelphia.

Pearsall, H. 2010. "From Brown to Green? Assessing Social Vulnerability to Environmental Gentrification in New York City." *Environment and Planning C: Government and Policy* 28(5), 872–86.

Philadelphia City Planning Commission. 1995. "Vacant Land in Philadelphia: A Report on Vacant Land Management and Neighborhood Restructuring." Philadelphia.

Pudup, M.B. 2008. "It Takes a Garden: Cultivating Citizen-Subjects in Organized Garden Projects." *Geoforum* 39(3): 1228–40. https://doi.org/10.1016/j.geoforum.2007.06.012.

Quastel, N. 2009. "Political Ecologies of Gentrification." *Urban Geography* 30(7): 694–725.

Reynolds, K. 2015. "Disparity Despite Diversity: Social Injustice in New York City's Urban Agriculture System." *Antipode* 47(1): 240–59.

Russ, Valerie. 2012. "Democracy 101: Callowhill NID Foes Went Up against Powerful Forces... and Won." *Philadelphia Daily News*, 6 February.

Schmelzkopf, Karen. 2002. "Incommensurability, Land Use, and the Right to Space: Community Gardens in New York City." *Urban Geography* 23(4): 323–43. https://doi.org/10.2747/0272-3638.23.4.323.

Seplow, Stephen. 1999. "Too Many Houses, Too Few Residents. S. Phila. Is Plagued by Abandoned Properties." *Philadelphia Inquirer*, 10 May.

Shinew, Kimberly J., Troy D. Glover, and Diana C. Parry. 2004. "Leisure Spaces as Potential Sites for Interracial Interaction: Community Gardens in Urban Areas." *Journal of Leisure Research* 36 (3): 336.

Slocum, Rachel. 2007. "Whiteness, Space, and Alternative Food Practice." *Geoforum* 38(3): 520–33. https://doi.org/10.1016/j.geoforum.2006.10.006.

Smith, C.M., and E.H. Kurtz. 2003. "Community Gardens and Politics of Scale in New York City." *Geographical Review* 93(2): 193–212.

Sokolove, M. 1994. "From Blight, Urban Oases Sprout: Gardeners Are Reclaiming Decayed Sites: A City Gardening Program Nurtures Them." *Philadelphia Inquirer*, 17 July.

Thibert, Joël. 2012. "Making Local Planning Work for Urban Agriculture in the North American Context: A View from the Ground." *Journal of Planning Education and Research* 32(3): 349–57. https://doi.org/10.1177/0739456X11431692.

Vitez, Michael. 1994. "Enter an Urban Garden of Eatin': A Plot, a Hoe, and Early Bird: Planting Season Has Begun." *Philadelphia Inquirer*, 24 March.

Vitiello, Domenic, and Michael Nairn. 2009. "Community Gardening in Philadelphia: 2008 Harvest Report." Philadelphia: Penn Planning and Urban Studies, University of Pennsylvania, http://www.farmlandinfo.org/sites/default/files/Philadelphia_Harvest_1.pdf

Voicu, Ioan, and Vicki Been. 2008. "The Effect of Community Gardens on Neighboring Property Values." *Real Estate Economics* 36(2): 241–83. https://doi.org/10.1111/j.1540-6229.2008.00213.x.

Wachter, Susan. 2004. *The Determinants of Neighborhood Transformation in Philadelphia: Identification and Analysis: The New Kensington Pilot Study.* Philadelphia: University of Pennsylvania, Wharton School.

Walker, R.E., C.R. Keane, and J.G. Burke. 2010. "Disparities and Access to Healthy Food in the United States: A Review of Food Deserts Literature." *Health and Place* 16(5): 876–84. https://doi.org/10.1016/j.healthplace.2010.04.013.

Wolch, J., J. Byrne, and J. Nowell. 2014. "Urban Green Space, Public Health, and Environmental Justice: The Challenge of Making Cities 'Just Green Enough.'" *Landscape and Urban Planning* 125 (2014): 234–44.

Acknowledgments

We would like to thank the many people who contributed to the development of this book, beginning with the growers, community members, and urban agricultural enthusiasts and sceptics who shared their experiences with us. We would also like to thank our editor, Doug Hildebrand, for his enthusiasm, guidance, and patience. The book benefited greatly from careful copy editing by Matthew Kudelka and the skilful attention of Anne Laughlin, the managing editor. Holly Knowles created the index and Korin Tangtrakul helped with preliminary research. Finally, we would like to acknowledge that this book was motivated by our students and their interest in urban agriculture and its transformative potential in neighbourhoods across Philadelphia. A special thanks to our colleagues at Temple University and to our families who have been supportive of the book and patient with us throughout this research and writing process. We especially appreciate Robert Mason's kindness and support.

Index

Printed and bound by CPI Group (UK) Ltd, Croydon, CR0 4YY

27/10/2024

14580702-0001